WHAT TEACHING TEEN MOMS TAUGHT ME

Janice Airhart

WHAT TEACHING TEEN MOMS TAUGHT ME

Lessons From a High School Classroom

Illustrations by

Tim Airhart

The Education Studies Collection

Collection Editor

Janise Hurtig

LPP

Dedicated to those of any age faced with sudden life changes:
You're capable of more than you think you are.

First published in 2025 by Lived Places Publishing

The author and editor have made every effort to ensure the accuracy of information contained in this publication but assume no responsibility for any errors, inaccuracies, inconsistencies, and omissions. Likewise, every effort has been made to contact copyright holders. If any copyright material has been reproduced unwittingly and without permission, the publisher will gladly receive information enabling them to rectify any error or omission in subsequent editions.

British Library Cataloguing in Publication Data
A CIP record for this book is available from the British Library.

ISBN: 9781917566001 (pbk)
ISBN: 9781917566025 (ePDF)
ISBN: 9781917566018 (ePUB)

The right of Janice Airhart to be identified as the Author of this work has been asserted by them in accordance with the Copyright, Design and Patents Act 1988.

Cover design by Fiachra McCarthy
Book design by Rachel Trolove of Twin Trail Design
Typeset by Newgen Publishing, UK

Lived Places Publishing
P.O. Box 1845
47 Echo Avenue
Miller Place, NY 11764

www.livedplacespublishing.com

Abstract

This book describes the experiences of a high school science teacher in a program dedicated to educating pregnant and parenting teens while supporting them with onsite childcare, nursing services, social services, and academic counseling. The teacher began teaching at age 55, without an education degree or certificate. She obtained an alternative secondary teaching certificate in science, based on her previous career as a medical laboratory scientist. During her eight years of teaching this special population, the author was surprised to find that her students taught her as much about life as she taught them about science. The book illustrates the many lessons she learned through stories and descriptions of situations unique to teen moms and their children on a high school campus designed to meet their needs. While these stories are provided in the context of science principles, each lesson is applicable to any teacher and any student population.

Keywords

Teen pregnancy; educating teen moms; alternative education; alternative teaching certificate; high school classroom; lesson planning; teaching strategies; educational philosophy; teacher-student relationship

Acknowledgments

I'm incredibly grateful for the students and staff at the Margaret Hudson Program with whom I worked between 2007 and 2015. Each of them contributed in some way to the most transformative journey of my life. Meanwhile, my teaching colleagues were an endless source of technical and emotional support. I'm humbled by the lessons I learned from them all. I would especially like to thank Linda Jones, Holly Martin, and Ashley Nickleson, as well as a few former students, for helping me remember details I'd forgotten.

For reading a draft version of the manuscript and providing comments, I'm indebted to Linda Jones, my former teaching mentor and colleague, whom I now consider a friend. Clara Snyder, the Public Services Librarian at the Leander Public Library, went above and beyond, passing on the manuscript to her partner and mother for comments after she'd read it.

Jan Hogle, Dawn Smith, Lisa Greinert, and Marty MacMillan, members of my various writing communities, gave me essential feedback on portions of the text. Thank you all for your insightful comments on content, construction, and mechanics. Most of all, thank you for your encouragement; it is priceless!

I also want to thank Rebecca Beardsall for editing an early draft of the manuscript with a keen eye and making suggestions that greatly improved the narrative. After several drafts and revisions,

Terra Friedman King provided thorough line and copy editing and proofreading.

My husband, Tim, deserves a shoutout for agreeing to sketch the wonderful illustrations throughout the book and for patiently reading and commenting on chapters in the vomit stage. His unconditional support, despite my sometimes harebrained ideas, has been vital.

Contents

Preface

Between 2007 and 2015, I had the privilege of interacting with hundreds of teen moms on a small campus where I began a third career as a high school science teacher at the age of 55. Many students were never enrolled in my classes but frequented my classroom almost daily. I also interacted often with my students' children, some of whom played on swings, slides, and tricycles just across the playground outside my classroom window. My days were full of brief, inconsequential exchanges. Though many of those conversations are etched clearly in my memory, it's impossible to recall exactly which student or child said or did what in every situation. For that reason, I've changed the names of students and babies throughout the book.

My years of teaching high school science were intense, and some details have escaped my memory. While I've done my best to describe events and conversations as completely and accurately as possible, there are likely incidents that others will remember differently. I encourage them to put their stories in writing as well. Mine is not the only story worth telling about the challenges of teen pregnancy or learning to become a teacher.

Learning objectives

1. Identify significant educational challenges faced by pregnant and parenting students.

2. Describe the elements of a support system that could assist teen moms in simultaneously overcoming educational challenges and becoming responsible parents.

3. Compare and contrast the effectiveness of standard curriculum requirements with uniquely designed teaching strategies in meeting the needs of special populations.

4. Assess the value of establishing trust in forming beneficial teacher-student relationships.

5. Describe how teachers and students learn from each other.

Part I
2007

Change

The most difficult thing is the decision to act, the rest is merely tenacity.

– Amelia Earhart

Part I Introduction: My entry into the teaching profession

Between 2007 and 2015, I taught science to teen moms at the Margaret Hudson Program (MHP), a small high school campus in Broken Arrow, Oklahoma, a suburb of Tulsa. My time, energy, and emotions were dedicated to the MHP mission of helping students succeed. The program served pregnant and parenting teens from area school districts, who could enroll as soon as they learned they were pregnant and stay for up to two years or until they graduated. In addition to extra academic support, onsite child care, parenting programs, and counseling supported students in becoming responsible mothers. I was just the science teacher, but my colleagues and I learned to become advocates for teenagers whose lives were upended by unexpected changes and the challenges of an unplanned pregnancy before completing high school.

I did not have a degree in education when I began teaching in 2007, at the age of 55. Instead, I had degrees in biology and journalism, with decades of experience in medical laboratory science and writing or editing. It didn't occur to me at the time that an education degree or teaching certificate could offer a better foundation than expertise in the subjects I taught. Within weeks, I realized how misguided my optimism was. While I had spent months taking credentialing exams to become provisionally certified and studying textbooks to plan lessons, I had not studied any texts on educational theory or classroom management. When most of the students in one class failed their first major exam, I knew I had failed to teach them well. Recognizing my failure, I consulted my teaching mentor, who provided much needed wisdom and encouragement. This was the beginning of my education as a teacher.

1
The stakes: Challenges facing teen moms

In the 1950s and 1960s, families often dealt with teen pregnancy by shipping the girl off to a relative in another city or state until her baby was born. There were also homes for "unwed mothers" if a relative could not be located or convinced to take the girl in. My own sister spent the latter months of her first pregnancy in Baton Rouge, Louisiana, several hours from our home in Lake Charles. My stepmother offered me this option when I became pregnant in 1971, but I declined. By that time, I was a working college student and chose to marry instead. Becoming parents at 19 was not easy for my husband and me, to say the least. On the other hand, it taught me a lot about perseverance and seeing the possibilities in unexpected situations. It was this lesson that I hoped to pass on to the girls in my classes, but I soon realized I had a lot to learn about who these teen girls were and how they ended up in my classroom. They weren't just statistics, they weren't just like me, and they weren't all the same.

Several factors affect birth rates and the ages at which pregnancy occurs. Living in poverty, the educational level of the mother, and the availability of birth control are the factors most relevant to global birth rates. As young women began attending college in the United States in higher numbers in the 1970s and 1980s, they found they had choices about employment versus homemaking. Childbearing was deliberately delayed. For teens at risk of unplanned pregnancy, other factors played a role. The advent of reliable birth control in the 1960s, and the availability of safe and legal abortion procedures beginning in 1973 decreased the teen birth rate as well. Fifty years later, the effects of the Supreme Court's action eliminating the universal right to abortion for US women are yet to be seen, but teen pregnancy rates may again rise.

A lot of dedicated energy by many people and organizations has been expended to promote wise choices about sex among teens in recent years, at least in those states willing to confront the issues directly. Those efforts are paying off and deserve celebration. Sex education in public schools is still not universal, however. Leaders in some states and many of their constituents are convinced that teaching about sex encourages sexual behavior, despite data proving the opposite. Instead, in states like Oklahoma, where I taught, teaching abstinence is the only requirement of the scattered and optional sex education classes. Consequently, teen pregnancy rates remain high. The support of programs like MHP, dedicated to educating teen moms, provided a vital role in helping young families succeed. Unfortunately, with the number of teen moms in decline, those programs have been shuttered. MHP closed permanently in 2017 because of dwindling financial support from the community.

Today in Oklahoma, a relatively small state, there are nearly 7,000 pregnant girls between the ages of 15 and 19. Will they complete their education? Many won't. This puts them at risk of a host of disadvantages. Nationally, the high school graduation rate was 87 percent in 2022. For teen moms, the average is around 50 percent. The disparity in these statistics suggests that traditional education and classrooms greatly shortchange teen moms. Because of the support MHP provided, the graduation rate for our students was greater than 90 percent.

High school dropouts are more likely than graduates to face long-term poverty and a long list of other disadvantages. If those dropouts are also parents, their children pay a steep price as well. We all pay a price when children, including children who become parents, don't reach their potential. I taught parenting students as young as 13.

Children of teen moms, half of whom are school dropouts, are at greater risk of many disadvantages. They risk low birth weight and infant mortality. They're less likely to be adequately educated and more likely to continue living in poverty and perpetuate that condition as they have families of their own, often as adolescents as well. They're more likely to be incarcerated during adolescence. A high percentage of my students were born to teen mothers.

The promise of making a difference in the lives of these young women and their babies was compelling. I was not, and still am not, confident that their best interests would be served by current educational and political systems. Instead, the issues of teen pregnancy, sex education, abortion, and birth control access become more fraught each year, ebbing and flowing with election cycles. Whether I had the resources necessary to significantly affect their futures was a question I barely considered in 2007, but I felt drawn to their circumstances. With the best of intentions, I stepped into the fray, eager to improve the lives of these girls.

MHP offered state-accredited childcare to 20 infants and toddlers onsite that consistently rated in the top tier. The teachers in the childcare department did more than care for children, though. They mentored and taught students how to care for their babies by demonstrating healthy mothering behaviors. Our nursing services provided well-baby checks, and counselors did much more than academic counseling. All these social services operated under the nonprofit arm of MHP, a United Way Agency, separate from the education services the local school district provided. They existed in a wing of the building just beyond the cafeteria. A Women, Infants, and Children (WIC) office was housed there as well to ensure moms' and babies' nutritional needs were met. Just down the block was a Head Start, early childhood learning site for children who aged out of our childcare, and a Planned Parenthood office. We invited community organizations like the Tulsa Public Library, Parents as Teachers, Junior Achievement, the Chamber of Commerce, and others to provide programs for our students. All our services were essential in helping the girls graduate and take their first steps into the adult world as responsible parents.

2
Teen pregnancy by the numbers*

- US teen birth rate steadily declined between 1991 and 2021 by 78 percent but is still higher than most high-income nations.
- In 2021, the US teen pregnancy rate on average was 1.44 percent.
- Rates vary considerably among racial and ethnic disparities, with the highest rates among non-Hispanic American Indian/Alaska Native girls at 2.4 percent and the lowest at 0.2 percent for non-Hispanic Asians.
- Pregnancy rate in older teens (18–19) was 2.73 percent in 2021.
- Approximately 15 percent of teen moms will have at least a second child while still a teen.
- In 2017, an estimated 58–60 percent of pregnancies resulted in live births and 23–28 percent ended in abortion.
- Oklahoma, where I taught, had the fifth highest teen pregnancy rate in the United States in 2022, at 2.12 percent. Mississippi had the highest rate at 2.64 percent, and Connecticut had the lowest rate at 0.64 percent.
- Among mothers who have had a child prior to age 18, only 38 percent have a high school diploma.

- Less than 2 percent of teens who become mothers before age 18 obtain a college degree before age 30.
- Eight out of ten teen dads don't marry the mother of their child.
- Teen girls who are sexually active but don't use contraceptives have a 90 percent chance of becoming pregnant in one year.

*Unless otherwise noted, the above US statistics apply to girls aged 15–19 years.

3
Birthday girl:
A new career at 55

"Here," Holly said. "Put this on." She thrust a laminated paper pin with a tag that said, "Birthday Girl" in my direction.

"What is it?" I'm not sure what look I gave her, but I hope I looked cooperative, like a respectable member of the teaching team I was joining.

"It's the birthday pin. Everyone wears it on their birthday." Holly taught Family and Consumer Science down at the end of the hall. "Perfect!" My new colleague stood back to assess the overall effect and patted me on the arm. "You'll be fine," she said.

Oh my God! I might as well have a scarlet "A" emblazoned on my chest. The pink and yellow birthday ribbon stood out like pulsing neon against my blouse.

"Thanks," I said weakly. I didn't protest; I *am* a team player, after all. *Everyone* wears the pin on their birthday. But I was hardly a girl.

***

It was 2007, my 55th birthday, and my first day as a high school science teacher at the Margaret Hudson Program. I dragged a wheeled cart behind me filled with everything I thought I'd need

for my teaching debut: four science textbooks and a spiral-bound lesson plan book complete with four class lists neatly printed in pencil (I was warned to expect frequent changes), worksheet masters, my purse, and a frozen entrée for lunch. Wedged among it all was a sense of optimism for this unlikely career change. The cart was so heavy that I worried I'd pull my shoulder out of joint hoisting it from the back seat of my car to the ground. What there wasn't room for was self-confidence in what I was about to do; it was too dodgy to pin down. There also was no eight-by-ten framed education degree certificate with my name on it. I was relying on my degree in biology, which the state of Oklahoma deemed sufficient, along with several satisfactory certificate test scores to teach science.

I'd been preparing for this day for weeks: reading science text-books, studying flow diagrams, planning detailed lessons, and setting up my new classroom. As the half-time science teacher, I'd share the room with the half-time math teacher, Gerald. Each of us represented one-fifth of the entire teaching staff on a campus with four classrooms. For days, I'd been stapling colorful posters and pictures on the single classroom bulletin board at the front of the room. Gerald wasn't so keen on decorating.

"Go ahead and put up whatever you like," Gerald said with a shrug from his end of the classroom.

Meanwhile, at my own desk, I carefully hung file folders with color-coded stickers, inserted a black mesh organizer in the long shallow drawer, and labeled and organized office supplies. Trying to take advantage of my cohort's vast experience, I show-ered him with questions. "How do you keep track of absences?"

I asked. "How do you know when to back up and cover a concept again?" He was patient with my questions and answered as calmly as I imagined he did with students. "I know girls are allowed to breastfeed in class. How does that work?" Not all my questions had to do with classroom procedure or with science, though that was the subject I was hired to teach. I was nervous about how I'd be perceived as a middle-aged, first-year, inexperienced teacher by my teenage students. I was older than their mothers, maybe as old as their grandmothers.

***

When I was first introduced to the Margaret Hudson Program (MHP) in 1989, it was housed in the basement of a Baptist church in a large suburb of Tulsa, where we'd recently moved. I signed up as a substitute teacher for Broken Arrow Public Schools while I looked for permanent work as a medical laboratory or research scientist, my profession for more than 20 years. The only substitute job I accepted before I found full-time work was subbing for the half-time math teacher at MHP.

I was impressed that Broken Arrow had a program dedicated to educating pregnant teens and caring for their children, the only one of its kind I'd ever encountered. I'd been pregnant myself at 18 as a college freshman. I never needed this kind of support, but I had great empathy for the girls who did. The school was jointly operated by the Margaret Hudson Program nonprofit agency, which was funded primarily by the United Way of Tulsa, and the Broken Arrow Public School District. The nonprofit organization provided nursing, counseling, and childcare staff—all of whom had degrees in their respective fields. The school district subsidized instructional staff and resources, including my position in 2007.

Several years after my short-lived substitute teaching career, and after I'd left my last job in biomedical research, I answered an appeal in the newspaper for mentors at MHP, the brainchild of a pediatrician and public health official named Margaret Hudson in Tulsa in the 1960s. She'd noticed the disadvantages faced by pregnant teens who mostly dropped out of school when they learned they were pregnant. She was instrumental in forging a partnership between the United Way and the local school district. I'd been intrigued by the program in 1989, and I wanted to learn more about it. I signed up to be a student mentor 12 years after substituting there.

Over the next three years, I worked with three different students with varying degrees of success. The last of the three girls left MHP, ostensibly to transfer back to her former high school mid-year, and I never knew if she graduated. It felt like a failure on my part, and I feared for her future. Pregnancy was—and still is—one of the most common reasons for girls to drop out of school.

***

As students filed into the classroom on that first day, my cheery smile was pinned as carefully to my lips as the "Birthday Girl" pin was on my chest, and I injected as jovial a lilt into my shaky voice as I could manage. I've never been more petrified.

"Sit wherever you like," I said with what I hoped was an encouraging smile and introduced myself as each girl came in for the first period. I'd arranged the two-person desks and blue plastic chairs in blocks of four, two facing two. Girls surveyed the room silently before choosing a seat. It was the quietest hour we would share in the classroom all year.

Students stared at my chest; some were discreet about it—others, not so much. "As you can see, it's my birthday today," I offered, pointing out the obvious. "What a special way to spend my day, too! Meeting all of you." In the quiet that followed, I launched into a brief introduction of myself. I told them I was married with two grown and married children and one young grandson. "I was a medical technologist for almost 25 years, working in several different hospital and research laboratories. I love science, and I hope you will too." Most avoided looking at me, or at anyone.

Our pregnant or parenting students came from several area school districts, but our campus wasn't large and accommodated only 50 girls. The majority transferred from the Broken Arrow District High School—the largest in Oklahoma—but we also served a few smaller districts in the area. Girls could enroll in our program as soon as they had a positive pregnancy test and could stay for a maximum of two years. Because our campus was small, no one remained anonymous. The building had two wings, perpendicular to a central hall which housed the staff breakroom, the administrator's office, and the reception area. These were across the hall from the cafeteria, the largest space on campus, where students and their babies ate meals and provided space for assemblies or school programs. One wing accommodated our small library and the four classrooms. The other wing, on the opposite end of the central hall, held social and nursing services, a childcare facility for 20 children from birth to 2, and a WIC (Women, Infants, and Children) office.

"I want us to get to know each other better," I said after I'd finished my introductory spiel. "Please introduce yourself to the class with your name and when your baby is due. Or if you're already a mom, tell us your baby's name and how old he or she is." Asking students to identify themselves in a class of girls they'd never seen before was probably the most frightening question I'd ask them all year, but I didn't want to be the only panicked person in the room.

I needn't have worried. What the girls had in common created bonds quickly. Some returning students already had infants or toddlers down the hall in childcare and became instant authorities on everything related to pregnancy and motherhood. The questions they asked each other made *me* blush. Within days, I couldn't shut them up. Within weeks, I dismantled the blocks of desks in favor of a large "U" that allowed me to circulate among them. Making eye contact over a desk, instead of speaking over heads or from behind them, became a valuable strategy for gaining attention.

***

There are a few maxims every teacher hears early in their career; I think I heard every single one. "Don't let them see you smile" seemed harsh. Its cousin, "Don't let them see you smile until Christmas" was a little better, but still a bit harsh. "Never let them see you sweat" was compatible with my general approach to new situations and had already proven itself useful. Hence, my pasted-on smile and upbeat speech. "You only need to stay one chapter ahead of your students." I was counting on this one. Teaching four different science subjects in three class periods, two of them simultaneously, meant I was already working as hard as I could to keep up. One chapter ahead was sometimes more than I could manage but provided a worthy goal. And then there was the impossible, "Relax and be yourself." The employee manual also cautioned against overly personal interactions between teachers and students. I took their admonition seriously. Perhaps too seriously in the beginning.

The principal, JoAnn, advised teachers to begin the semester by going over classroom policies and procedures. I drafted a lengthy document, including rules and the consequences of breaking them, based on a classroom management book I'd read. There were a couple of pages of how homework and tests would be graded, and how grades would be calculated. I tried to pare it down to essentials, but it came in at more than two thousand words anyway. It's still on my hard drive, and it makes me blush to read it today. I shouldn't have been surprised when the girls ignored it all. Even *I* was bored reading it. The students' lack of response should have provided an opportunity for enlightenment, but establishing policies was written in my lesson plan, so I plowed on.

Playing Get Acquainted Bingo afterward was a bit more popular, but I'd overestimated how much conversation the activity would generate. We had time to kill. An early lesson learned as a schoolteacher: Always have more material planned than you can possibly hope to cover. You never know when you'll need it.

I'm not sure how I expected that first day to go, but I think the picture in my head included a radiant glow from smiling faces in a roomful of grateful students and a warm conviction in my heart that I'd give these girls the benefit of my experiences and knowledge.

The only reaction I *am* certain of from my 55th birthday is my dismay at what I interpreted as disinterested students who didn't trust this weird, older science teacher who'd waltzed into their classroom with a cheery smile and a four-page list of expectations. This woman they knew no better than they knew their classmates or the cashier at the Taco Bell drive-through. This school they'd never expected to need. This pregnancy they hadn't planned on. Instead of feeling called to be there, as I did, every one of them wished they were anywhere but where they found themselves.

Naively oblivious to how stunned and terrified my students were to be teenagers and pregnant, and unaware that they were more frightened of me than I was of them, I pretended I had everything under control. Even if I'd felt their fear, I wouldn't have known what to do about it. I wasn't as flexible or as perceptive as I thought I was, though in time my students would teach me those skills. Who did I think I was to make this radical career change so late in life? It took me months to begin entertaining that question. On that first day as a classroom teacher much closer to a social security check than to my first professional paycheck, I stuck to my carefully crafted lesson plan. It was all I had.

Physical Science Test Framework

X SUBAREA I — SCIENCE PRACTICES
X Conduct scientific investigations
X Apply knowledge of methods + procedures
  Analyze the appropriateness of a
  specific experimental design

SUBAREA II - FORCES, ELECTROMAGNETICS, WAVES

Apply knowledge of Newton's Laws
of motion + solve related problems, including
X in relations to natural phenomenon (e.g. ball
rolling down a ramp, planetary motion
space walk)

SUBAREA III - MATTER AND ENERGY

SUBAREA IV - EARTH AND SPACE SYSTEMS

SUBAREA V - PEDAGOGICAL CONTENT KNOWLEDGE

4
Mission accepted: Meeting a significant community need

"That's ridiculous," my husband, Tim, said when I told him I planned to obtain an alternative teaching certificate to teach high school science. "Why would you do that?"

I can't blame him for his skepticism. It was totally out of character for me, a confirmed introvert without a teaching degree. On the strength of my master's degree in journalism, though, I'd begun teaching evening English classes to adults at Tulsa Community College a year earlier, and I found to my surprise that I enjoyed it.

"I read an article in the paper about the secondary school certification process," I told him. "They need high school teachers badly, especially in science. You don't have to have a degree in education to become certified."

Tim just shook his head, not yet convinced I'd go through with what he considered a nonsensical career change. Perhaps he was just afraid any argument would propel me to dig in with greater conviction. I could be like that. What neither of us knew at the time was how crucial he would become in supporting me, both physically and emotionally. If I needed a whiteboard moved to another wall in my classroom, a balcony constructed for our gerbil cage, the loan of a specific tool from his collection, or a lesson on how to use it, he was my go-to. He's still my best support system.

Our son was a bit more encouraging at the outset. "Gee, Mom. I never know *what* you'll do next." I chose to believe he meant my goal was inspiring, but I didn't ask him to elaborate. Our daughter accepted my announcement without much comment, aside from asking what was required to get certified. Maybe she assumed I knew what I was doing.

Besides a regular paycheck and full health insurance coverage, I was eager to find a job where I felt useful. My mother was diagnosed with schizophrenia shortly after my birth and institutionalized until her death when I was 13. Growing up motherless and yearning for a mother's presence in my life led me toward working to improve situations for vulnerable children. I had mentored a girl in foster care for several years in the late 1990s and volunteered with child welfare departments in several states prior to that. In the early 2000s, I'd been a volunteer student mentor at MHP, where two generations of children—moms and babies—struggled to make the best of a difficult situation. Improving their circumstances appealed to me.

***

The rules of the physical science certification exam I was required to pass, printed in minuscule font, filled both sides of the "Candidate Rules Agreement" I'd signed just to gain admission to the noisy, crowded high school auditorium lobby at 7:30 a.m. sharp on this Saturday morning in July 2007. Doors would be locked at 7:50.

The rules stated:

> Cell phones are prohibited on the premises … Jewelry, watches, wallets, and purses must be stored outside the testing area … Barrettes or hair clips larger than ¼ inch (½ centimeter) wide, headbands larger than ½ inch (1 centimeter) wide, or hats (and other non-religious head coverings) are not allowed … A scan of your palm vein pattern may also be collected.

Holy crap! Imagining how people might cheat with barrettes or hair clips was intriguing, but I was mostly annoyed that cheaters made taking an honest exam tricky for the rest of us. Despite feeling I was well prepared, the palms they might deem necessary to scan (why, I still have no idea) were damp. I couldn't stand still. If I'd been free to pace, it might've calmed me, but I didn't dare give up my place in line. I couldn't afford to be disqualified. The job I'd recently accepted was more than just a job to me. It was a position that would give me a chance to benefit girls for whom I had great empathy, a goal consistent with much of the volunteer work I'd done for years. It gave me a vital sense of purpose.

Once inside the test site, I'd have four hours to prove what I knew about physical science in 80 questions and one essay. I'd need 240 out of 300 total points to get the required teaching certificate. It was mid-July and school started on August 9, my birthday.

I'd borrowed a copy of the physical science textbook off the MHP classroom shelf back in May. This was right after the principal,

JoAnn, offered me a half-time job teaching four different science courses in three class periods every morning.

JoAnn remembered me from my time as a student mentor and offered me the position after talking with me for just a few minutes. It didn't feel much like an interview, really. She knew I was dependable and cared about the students. I think she just wanted to get the position filled. She called me later that day to say she shouldn't have offered me the job on the spot. "I goofed. I should have sent you to Human Resources first," she said. "But I smoothed it over. They'll be calling you to fill out some paperwork."

Just like that, I was hired. But there was one hitch. My biology certificate would allow me to teach biology, environmental science, and anatomy, but I would need to pass the physical science exam before teaching that class. It was already June.

For the past several weeks, I had studied borrowed texts, science tomes from the used bookstore, and AP physics and chemistry reviews—physics and chemistry being the two sciences that make up ninth-grade physical science. I created a grid of topics from the state-issued study guide and marked my progress with "X" marks.

Finally seated with my exam booklet and blank answer sheet, I aligned three freshly sharpened pencils with virgin pink erasers along one edge of the desk arm that pulled up around me and pinned me to my seat. At the test proctor's instruction, I filled in the bubbles corresponding to the letters in my name and the identifying numbers from my ticket, going over and over the circles to ensure they were graphite black.

In keeping with my analytical nature, I drafted a five-step test-taking strategy in my study notes. I mentally reviewed the steps and began with the first one: *Scan the entire test to consider how much time to spend on each section.*

My body began to relax. The questions all seemed familiar. *I can do this,* I thought and began reading and filling in circles, guided by the remaining steps of my plan: (*1*) *Read each question and possible responses twice, (2) After filling in a circle fully, doublecheck I've filled the correct bubble, (3) If I'm not sure of an answer, skip that question and come back to it later, (4) Create an outline of the essay response before writing.*

My strategy was foolproof. Hadn't I already passed the basic qualifying and biology exams? Hadn't I just been offered a job teaching in the school I'd volunteered at where I now pictured myself surrounded by eager young students? The school's mission statement was "Helping Teen Families Succeed." After raising two children and making the most of my own educational and employment opportunities, I could share hard-won wisdom with these young moms to help them succeed. I felt a call—an *obligation* almost—to be their teacher.

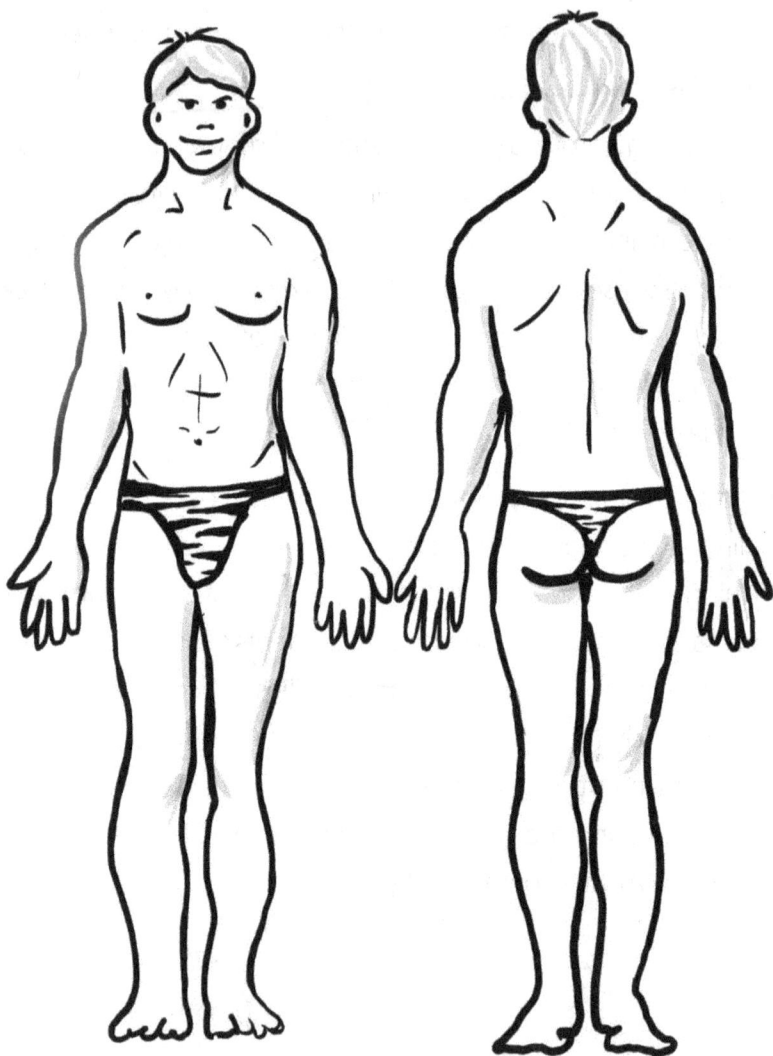

5
Anatomical position: Face forward, confidently

One of the first lessons in *Hole's Essentials of Human Anatomy and Physiology* is a description of what is called "anatomical position," the standard pose used in medical illustrations. The photo in our textbook showed a man standing straight, facing forward, with arms slightly extended from the sides and palms open and forward. A second image showed the man in the same pose from the rear, or dorsal view. From this stance, relative positions for body structures are derived and form the basis of understanding where in the human body a system or process is located.

Anatomical position implies a certain physical strength, as though the model in the photograph were facing his struggles head-on and with an open posture. *Bring it on!* During my first week at MHP, I took my anatomy students' photographs in anatomical position. After I printed the students' photos, they dissected them and marked imaginary body quadrants and relative positions. I still have the students' photos. While one girl smiled confidently for her photo like the man in the text, one looked bored and the others looked terrified. I hoped students would better connect to the lesson if they could picture themselves standing in that position, and perhaps they did. Perhaps it was early enough in the year they didn't yet know what to expect from me. Maybe I should've had them take my photo, too. I wonder how I appeared to them.

Without a doubt, Anatomy and Physiology (A&P) proved the most challenging course to teach, especially with only a handful of students. However, it was also the course I most looked forward to. After having spent so many years employed in medical settings, I was fascinated by the human body. I was keen to ignite students' passion about it as well. Surely these girls whose amazing bodies could bring forth new life would share my wonder at how organ systems work together to sustain a developing human being.

The A&P class roll I printed about a week after school started in August 2007 included eight students, but in the first few weeks of a semester, our rosters changed more frequently than a toddler changes her pull-ups. Somehow, there were only four students captured in anatomical position that year.

More than once, I was given the option to drop A&P since it was least preferred by students, but I insisted on keeping it on my course load because *I* preferred it. The previous science teacher had taught four science courses in three class periods. Two subjects were taught simultaneously during one of those periods, and I was expected to do the same. I chose not to drop A&P, determined to prove I was as competent as my predecessor, not realizing just how much this decision would complicate my life.

Teaching two courses in the same classroom at the same time seemed an invigorating challenge until I tried it. It felt like I was cheating both classes because I could only devote half the period to each. When given independent work, many didn't know what to do and consequently did little. When the bell rang, I barely caught my breath before the classroom emptied and filled again. There wasn't time to visit the bathroom until lunchtime. I was always *on*—on stage, so to speak, which is a model of teaching that I'd been advised against. To combat the "sage on the stage" phenomenon, I spent my weekends researching and planning more interesting activities to fit each unit.

I dug into print and online resources to discover startling facts and eagerly looked forward to my students' reception. I wanted them to love the quirky, surprising, sometimes unbelievable world of science, but despite all my efforts in those first months of class, students didn't seem enthused. The fact that they weren't dazzled by it confused me.

In the beginning, I dismissed the possibility that I was part of the problem. One realization grew slowly over the first few weeks: I knew science, but I had no idea how to teach it. I'd been so intent the previous summer on studying to pass the subject matter exams and obtain certification that I'd given no thought to how I would accomplish the actual transfer of knowledge tucked into my brain cells. At first, I assigned daily reading from the textbook or rehashed chapters aloud during class times, but I soon saw that wasn't teaching. While the students seemed to respect that

I knew what I was talking about, they appeared to have little desire to learn it. Perhaps they concluded they didn't have the capacity or that it required too much energy. Perhaps the students' changing physical and emotional circumstances left little energy to expend. Perhaps it was simply a dull presentation.

One of my A&P students even offered some advice during one particularly boring day of discussion—meaning I'd been asking questions about the text and answering them myself in the silence that followed. I was frustrated by their lack of response, making me even more anxious. While I was determined not to let students see me flounder, they'd clearly seen through me.

"Maybe you all need to read that section again before we talk about it," is what I remember stating with a great deal of irritation. "Read it again," I said.

"Maybe you could give us some diagrams to label or activities to help us remember what's in the book," Chloe said. It felt like an accusation.

"Maybe," I said with a nod. Something about her suggestion both shamed and motivated me.

I still remember the two worst teachers I'd had—one in high school and one in college. History was never a favorite subject, but Mr Anderson made me hate it. He was a basketball coach first, a US history teacher last. He seemed to think if we could memorize every significant date in American history and recite every president in order, we'd learn history. The pattern of each unit was the same: read the chapter, answer the questions at the end, and take a test. Read the next chapter and repeat. Class discussions consisted mostly of being quizzed about the chapter and answering aloud the questions we'd already answered and turned in. I was never so bored in my life. Until my 8:00 a.m. parasitology class in college, when I'd awaken only minutes before a hurried breakfast and frenzied dash to class. The instructor basically read the textbook aloud in much the same droning monotone as a sleep meditation. It was the only class I ever dozed off in.

I knew neither strategy was authentic teaching—how had I become like them? I was passionate about science, but I was clueless in generating that same interest in my students. They were clearly not engaged by my mind-numbing reviews of the textbook, which I spiced up by projecting the text illustrations on my whiteboard—my first creative instinct. I pointed out the location of each organ or body part being addressed, and it seemed they barely looked up.

In mid-September, I administered the first chapter test. In the review session the day before the exam, there was little participation or discussion. I ended up answering most of my own questions in between nervous silences. By test day, I just wanted it over with so we could move on to chapter two. Surely they would be more engaged in that material, I thought.

As soon as my A&P class ended, I eagerly graded the chapter one tests, hopeful that grades would demonstrate better student understanding than I'd observed in class. Maybe the girls were bored with my presentation because the topics were transparently easy.

It was soon clear that wasn't the case. I was stunned. What had gone wrong? I sat in the quiet classroom for a few more minutes, near tears. As required by the state's alternative certification guidelines, I'd been assigned a teaching mentor who taught English and technology classes across the hall. I'd been at MHP long enough to recognize the genuine respect students felt for Linda, and I hoped to benefit from her experience. At the same time, I'd so far resisted the urge to burden her with my insecurities. Maybe I was just ashamed of how poorly I was performing. I finally decided to swallow my pride and walked across the hall.

"Do you have a minute?"

In addition to teaching English and technology courses full time, Linda was Lead Teacher and Instructional Technology Coordinator for our small campus. I hated to be a bother. Linda set aside what she was working on and motioned for me to sit down at one of the student desks. She pulled up a chair next to me. "What's up? You look upset."

"I just gave my A&P class their first chapter test." My voice was shaky. "It was bad."

"How did they do?"

"Almost all of them failed," I told her.

"I see. That *is* bad. What do you think went wrong?"

"I wish I knew." I was fully aware of having taken this position after the longtime science teacher recently transferred to the district's alternative campus. I also knew her personally. When Cindy and I talked about my taking on this job after her, she tried to be helpful. "Go with the flow," she'd advised, or something like it. At that time, it sounded reasonable, but an unspecified flow seemed easier to go with than the particular flow I was currently facing.

"I expect Cindy could make the class seem fun. I'm sure she could get the girls to talk and ask questions. I'm just not spontaneous like she is," I told Linda. "That's not me." Tears sprang to my eyes. "I don't know what to do. The girls failed the test, but somehow, I feel like *I've* failed."

Linda passed me a tissue and made a few soothing comments while I dabbed at my eyes.

I attempted a wobbly grin and sighed. "This is harder than I thought."

Linda smiled. "It *is* hard," she said. "Why don't I sit in on your class tomorrow while you review the test with students? Maybe observing how they respond will give me some ideas."

***

I spent more time than usual that evening preparing for the exam review. In the morning, I photocopied digital images that came with the curriculum onto transparencies for the overhead projector that teachers shared, and I arranged to use the projector.

As soon as students entered the classroom and saw Linda sitting in the back, the tenor of the room changed. They spoke to her, but somewhat nervously. Perhaps they were worried she was there to observe their behavior. They were uneasy, and I suspect they thought they were in trouble. Maybe they were afraid to see their exam grades.

For approximately 54 minutes, and for the first time, I glimpsed a different possibility for my future in the classroom. Students responded to questions … accurately, for the most part. They opened their textbooks to the appropriate pages. They volunteered comments and seemed engaged by content in the way I had imagined they would. I was aware that the difference was Linda. This was a teacher that students both admired and respected. She knew each student and each student's situation. She genuinely cared about them and wanted them to succeed. They knew all this about her, and that made all the difference. I remember thinking *I want to be like this someday.*

Linda's presence at the back of the classroom gave me confidence too. I loosened up a little and could feel myself relax.

I pointed to my transparency slides and asked, "Which body region is this? Check the illustration in your book if you're not sure."

When the answer chorused back from the whole class, "Right Lumbar!" I could feel myself smile. There was a spark there.

"What about this one?" A thrill coursed through me at their correct response.

Linda sneaked out of the classroom about halfway through the class period, but amazingly, students continued to interact. I told them I planned to administer the same test again the following day. "I want you all to succeed in this class." I could almost hear the tension release, like air swishing out of a taut balloon.

After class, I skipped across the hall. "Can you come to my class *every* day?" I asked Linda with a smirk. "I've never heard so many words come out of the mouths of my students. At least, not about anatomy. Why didn't they put all this effort into their tests?" I shook my head. "Who are these aliens, and what if yesterday's class comes back tomorrow and discovers they've been hijacked?"

Linda just laughed. "I think you'll be fine," she said. "They will too."

I'd like to say everything changed after that day, but it didn't. First impressions of a stuffy, uninspiring teacher are challenging to overcome, and it takes time to develop student trust. But the day my class failed their test and made me cry opened the door to learning what I needed to become a teacher.

Now I knew I'd need to discover better ways to demonstrate beloved concepts if I expected anyone else to share in my delight. If I wanted to stand confidently before students, facing forward, arms extended and open to what they needed, like the man who modeled anatomical position in our text, I'd have to first learn to assume a self-assured *teacher* position.

Part II
2007–2015

Learn

Tell me and I forget; teach me and I may remember; involve me and I learn.

—attributed to Benjamin Franklin

Part II Introduction: Importance of remaining open to change and to learning

I began my teaching position at MHP believing I had some wisdom to offer my students. As a former teen mom myself, I parented—with my husband—two children who became independent, responsible, and well-adjusted adults. At the same time, I had a postgraduate degree and a deep knowledge of science. These would surely all be assets in my new career. What I had not considered was that while I was hired to teach the students, my students would also teach me.

Each of the following chapters describes a lesson I learned from students, told through stories or student interactions. Because I was first a scientist and then a science teacher, I chose to frame the chapters in the context of science concepts. The concepts relate to the challenges faced by my students or to the individual care provided by MHP to help them succeed as both students and as mothers. At this intersection between my students and me and the academic support they received through a special program designed just for them, I became a teacher and an advocate for them. While the girls learned to adapt to an unplanned pregnancy and the challenges of motherhood, I learned to trust and respect their goals. I discovered that my greatest asset as a teacher was humility.

Their Stories: Why I got pregnant.

~ Top Ten ~

1. He said he loved me.
2. My parents wouldn't let me get birth control. They said it would encourage me to have sex.
3. My mom found out I was having sex and she said I should stop. I didn't.
4. My mom wanted to be a grandmother. Or, my Mom just LOVES babies!
5. My boyfriend's mother wanted us to get married so I'd move in and take over the housekeeping.
6. After your Quinceañera, you're a woman.
7. My boyfriend didn't want me looking at other boys.
8. I didn't know how to say "no" without hurting his feelings.
9. I like having sex.
10. He bought me a mega-ride pass to the state fair.

BONUS: "Oops!"

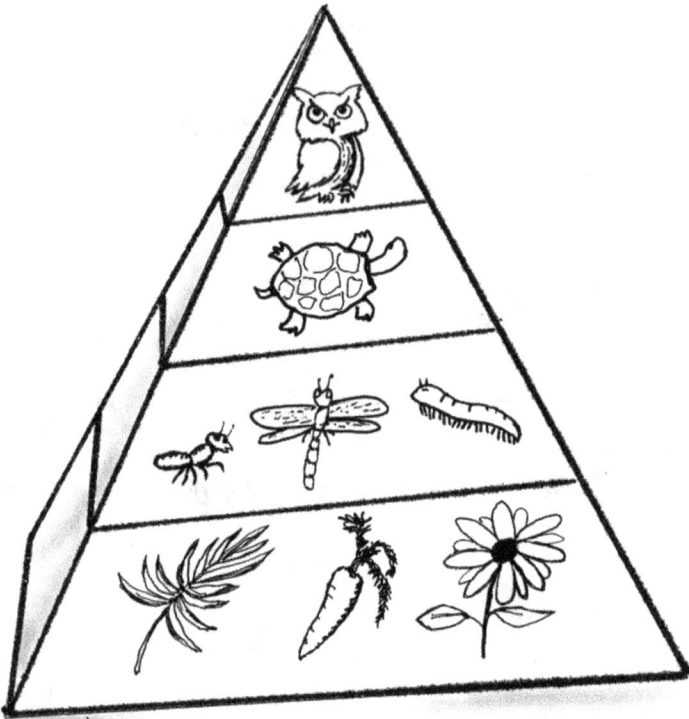

Energy Pyramid

6
Scientific method: Failure is an excellent teacher

"What do you see?" It was a simple question, but it was met with quiet squirming. Some looked at me with brows gathered.

Finally, someone ventured an answer. "It's a candle." I could almost hear the "Duh!" she was thinking.

"Yes. But tell me more. What color is it? What shape is it?"

After the girls had described the candle in detail, I pulled out a pack of matches and lit the candle. "Now tell me what you see." This was the first lab I conducted in all my science classes. By the time it was done, the girls had filled a page with observations about the candle, both lit and unlit. They'd recorded physical characteristics and how the wax melted and dripped down the side as the wick burned. They recorded smells and the warmth they felt when they held their hands near the flame.

"Remember how to really look at things when we do our labs this year. Observe what you see and hear, how things feel or smell. Taste, even." Some of the girls grimaced.

***

The first year I taught high school science, I stapled a large poster on the bulletin board that listed, in colorful, attractive fonts and graphics, the steps in the Scientific Method. After my years as a lab scientist and quite a bit of experience with labs gone wrong—often due to operator error—I wasn't altogether sold on the process as a simplistic, step-by-step procedure. It was a reasonable starting point, though. While I think the order of the steps should be more flexible than listed, they're still valuable moves in problem-solving. Teaching the steps seemed a useful exercise since my habit is to approach most problems scientifically. I hoped my students would find them useful too.

During those first couple of days of the semester, I led all four classes through a few simple experiments in demonstrating the following actions:

- Observe and ask questions about your observations.
- Research the topic to understand the variables.
- Form a testable hypothesis, or prediction, that answers the main question.
- Design and complete an experiment that tests the hypothesis.
- Collect and analyze data, using charts or graphs as necessary.
- Draw and report conclusions by answering the original question.

I'm not sure what exposure my ninth and tenth graders had to the scientific method before coming to MHP, but I sensed experimental processes were unfamiliar. In time, I realized many students had rarely done actual experiments due to the size of classes and perennially small budgets at the typical middle or high school. I thought it a priority in understanding science, however. My students had other priorities: false labor contractions, painful breastfeeding, boyfriend woes, or sleep deprivation, for instance. While I acknowledged our different priorities, I persisted in the notion that learning science would be valuable to them.

I planned labs and demonstrations to lead students through each step of the scientific method, beginning with the observation of a lit candle to describe changes over time and teach them how to attend to details. While attention to detail has a great many applications, the point in terms of science class was to assess if labs were working as designed and to adjust when they weren't.

A written lab report was necessary so I could grade their understanding of our activity, but its accuracy wasn't as important as their developing expertise in conducting experiments. What was built in as often as possible, even in year one, was the hands-on experience itself—the cornerstone of science teaching in every field. I loved labs!

My students loved labs, too, but just the movement from desks to the lab area could throw them off track. The science lab consisted of a niche in the back end of the classroom where there was a large rolling lab table with a traditional black resin top. It was equipped with an upright steel bar to clamp tubes or flasks, electrical outlets, and a small stainless sink with a spigot that drew water from a large plastic reservoir underneath. This is where most of our labs were conducted.

Some of my classes were quite small—only three or four—so we often just worked in one group. No matter how they were structured, labs gave rise to a certain amount of disorder. I felt it was my duty to maintain control, which was a frustrating exercise. I can see now that I took orderliness too seriously. I eventually learned to lighten up, but in the early years, I was dismayed by the side talk. One of the foundational labs, "Egg Floats" tested which combination of water and dissolved solid would allow the egg to float instead of sink. The experiment might go something like this.

"Mrs Airhart, did you know Lisa had to stop breastfeeding because of a boob infection?"

"Hmm. No, I didn't." I refined some answers to the fewest possible words. "Does everyone have a lab sheet? Be sure you have something to write with." This invariably led to a search for pens and pencils. (In year two, I kept a box of sharpened pencils near the lab table.)

"Did you see The Bachelor last night? Isn't Devon to die for?"

"Yeah, I hope he sees through that Heather. She's a skank."

"All right, girls. Let's fill out your predictions on your lab sheets." Side conversations sprang up faster than leaks from a punctured water hose. My challenge was to help them refocus.

"What should we put for 'starting observation'?"

Before I could respond, one girl might lean over to her neighbor and whisper loudly. "I hope Devon chooses Sherry. She's just the cutest thing."

Someone else might offer a relevant observation. "The eggs are cold. Does that make a difference?"

"Great question. Write that on your sheet." Interruptions were inevitable. I ignored what I could and moved on, opening the carton of eggs. "What do we know about eggs in water? Do they float or sink?" I'd pass the carton around so each girl could take one.

"I heard they float if they're rotten."

"I just bought these yesterday, so I doubt they're rotten. Let's see what happens when we put them in this big beaker of water." With a bit of prodding and well-timed suggestions, we managed. The lessons I planned for the rest of the week would hinge on understanding how to structure a hypothesis, so I reviewed the lab process again before the next activity. If I'd only been teaching one course, I could have spent more time taking useful notes in my lesson plans for the following year, but with four preps for four science classes, it was challenging enough just to prepare for next week.

Despite my typical scientific attitude toward solving problems, it didn't occur to me at first to apply the Scientific Method to the frustrations I experienced in the classroom. Even if I had, I'm not sure I would have had time to employ enough experimentation for it to be effective. In each succeeding year, I tried a slightly different approach to challenging concepts. At that rate, I figured I'd be 94 before I grasped the best teaching strategy for any course unit.

Like students describing what happened to a lit candle, I observed what was happening in the classroom while attached to my own priorities. Sometimes, the only question I could ask myself in the quiet after students left the classroom was, "What went wrong?" In that first year, I was stunned at times to realize that nothing I'd planned hit its mark.

At the same time, I excelled at writing lesson plans. Creating them was akin to designing an experiment. Being the consummate planner and organizer … that was my comfort zone. It was where I spent more than half my energy. It wasn't my fault I couldn't adequately predict how some variables would affect results. When a student's baby developed a respiratory virus and exposed the other babies in childcare, all of them then had to be sent home with their mothers, who were also my students. One of the mothers relied on her family's only car, which had been riddled with bullet holes over the weekend by her brother's rival gang members. This meant she was without transportation to school until the family could locate another beater car. Sometimes a student was overwrought by being forced to move in with her boyfriend's mother overnight when her father threw all her belongings—meager though they were—on the front lawn.

After prior experience with substitute teaching and mentoring at the school, I felt I knew the student body. The reality was that I didn't. Compounding the problem, I'd begun this job with the knowledge that I didn't have experience teaching this age group, but with the naive notion that I knew enough subject matter to transfer it to students.

By my second year of teaching, I'd learned a few things from prior mistakes and a few, mostly accidental successes. In other areas,

I adjusted activities to explain concepts via arts and crafts, for instance, or to demonstrate the differences between renewable and nonrenewable energy via heat currents in the closed system I built into an aquarium. I sometimes appreciated the creativity more than my students did, but at least I was keeping myself entertained.

A peculiar aspect of science methodology is that in analyzing results, the astute scientist doesn't accept a single instance of success or failure. Replication of results is mandatory to draw solid conclusions. That first year, I stumbled onto an energy pyramid activity, which used only cardstock, colored pencils, glue, and string. It was another simple and inexpensive activity, an important consideration. Our campus was allocated a pittance for supplies, each semester to split between our five faculty members; my portion of the pittance carried me through a month if I was lucky.

I approached the energy pyramid project as a craft activity with an instructional bonus. Every year, I set aside a couple of class periods to create the pyramids I later hung from the classroom ceiling like mobiles. Students both enjoyed the artistic nature of the assignment and better understood the concepts it was intended to teach after assembling them. While some variables of the class changed by virtue of student personalities, age range, or class size, this lab was replicated with confidence. Not all were.

Some labs were useless if I had only two students in a class, for instance. Others made more sense to moms who'd delivered their children than to those who were still pregnant. If some students were making up credits as seniors and the rest of the class were freshmen, the dynamics changed the tenor of the class entirely. Labs that weathered all these variables became a permanent fixture in my lesson planning toolbox, as this one did. By my third or fourth year, I'd honed several activities that enabled me to build on successes from the previous year. Of course, I first needed to have positive outcomes to replicate. "Keep this, discard that" was the order of those middle years, and through persistence, I began to gain ground as both a self-directed scientist and a teacher.

Even though the colorful scientific method poster was a rather simplistic attempt to introduce logical thinking to students, following the steps out of order sometimes made sense and kept me flexible. Whenever an observation in one class connected to a creative idea in my brain, progress was made in the next. I learned to allow for a few variables in a real classroom environment and not become too discouraged by less-than-perfect results. I studied my experiment design and revised it as needed. One scientific principle I explained frequently to students after every imperfect lab was one I finally had to accept myself: You learn just as much from your failures as you do from your successes. Sometimes more.

7
Atomic principles: Students can be teachers too

"Isn't it ingenious?" I asked. "The Periodic Table of the Elements is organized into rows and columns, and you can learn so much about the structure of an element just by looking at which row or column it occupies on the table." I could almost feel the goosebumps rise as I pointed out a few of those elements. "It's kind of a miracle that all matter on Earth can be so neatly organized, don't you think?"

From the skeptical expressions on their faces, it was clear my students didn't agree.

"Right, Mrs Airhart," Maria said with a nod.

For concrete thinkers, seeing is believing, and atoms can't be seen.

Let me just say this: that everything is made up of atoms and that atoms are made up of subatomic particles called protons, neutrons, and electrons—none of which can be seen—is a hard sell to 14-year-olds. Pretty much all of science rests on the principles of atomic structure, however. It was the first academic concept I approached each year in my ninth-grade physical science class.

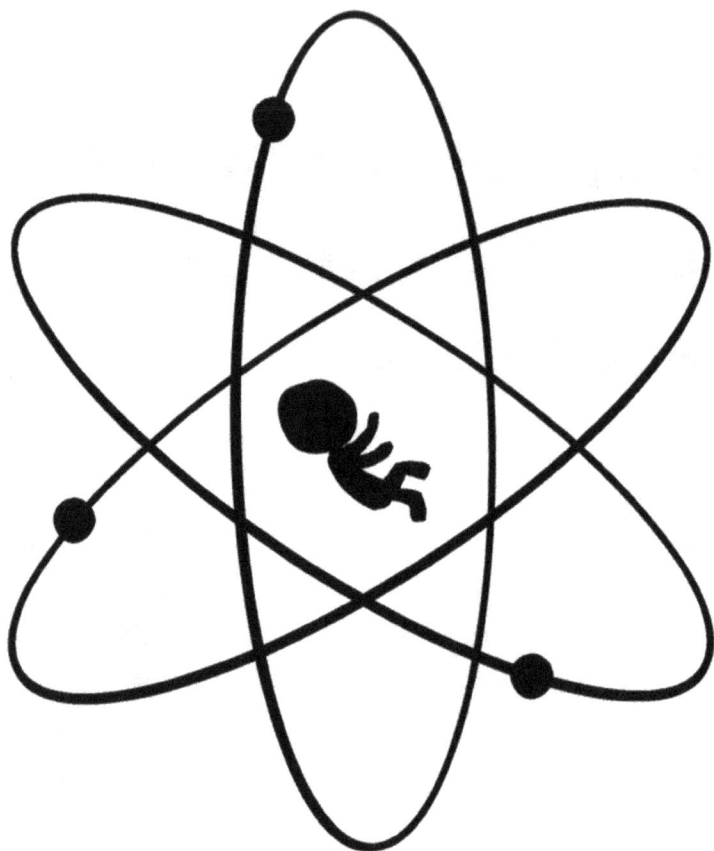

In theory, I could then move on to revealing the beauty of the periodic table and the characteristics and strengths of different chemical bonds. Getting my students to understand how critical the balance of negatively and positively charged particles is for an atomic structure to be stable was a challenge, and I kept reinventing strategies.

We constructed various atom models, like hydrogen or oxygen, using kits with parts that resembled Tinker Toys. Students stuck spheres together with little straw-like connectors that signified chemical bonds. Later in the year, we'd build molecules from these same spheres, but with different connectors to symbolize shared or donated electrons.

One year, we built playdough models of our atoms, and another year we created mobiles out of paper plates and pipe cleaners. I still have an ample supply of the tiny red, yellow, and blue dots we stuck on the nuclei and the orbits we cut from paper plates. The girls enjoyed the hands-on activities, but understanding what the project represented was still sometimes elusive.

Then there was the shower curtain exercise. I got the idea in one of the dozens of science idea books I'd purchased. I first spent the summer of 2008 decompressing from the anxiety of my first year in the classroom and then researching ideas online and combing through science education texts for activities illustrating difficult concepts. The shower curtain periodic table looked perfect, with blocks of variously colored paper representing all the elements taped into their proper rows and columns. The natural order of the periodic table soothed me, giving me the impression of regaining a bit of the control I felt I'd surrendered the previous year. Placing elements via paper blocks on the shower curtain, where each one had a logical place, based on the number of electrons (columns) and the energy level (rows), seemed to be a tangible way to organize scattered thoughts.

In the fall, I bought a white vinyl shower curtain and drew rectangles with the help of a T-square and a Sharpie to represent every currently known element. The columns and rows were arranged to mirror the colorful chart on the back cover of our textbook. Then I cut rectangles out of colored craft paper, in different colors for different element groups. On each paper rectangle was printed the labels and blank lines I asked students to fill in: atomic number, element name and symbol, atomic mass, and other data from the chart in their textbooks. I assigned each student a group of elements to create cards for and then passed the pieces of paper out to my students to complete by hand with permanent markers.

***

Georgia was nearing 19, the age limit for graduation from high school in Oklahoma. She was pregnant with her second child and had missed some earlier credits because of the birth of her first child. She was picking up her physical science credit before time ran out. Older than her classmates, she often attempted to mentor them and would sometimes explain a lesson in terms different from the ones I used.

Georgia's interruptions and addenda to my lessons could be irritating. Sometimes, they weren't factual. Other times, they disrupted the entire class period. Georgia had a "salty" vocabulary, to say the least, and she wasn't shy about describing her personal experience of labor and delivery to younger students.

"Labor lasted nearly 18 hours, and let me tell you, it hurt like hell," she'd say in the middle of an explanation of how many electrons an energy level could accommodate. With a glance at my perturbed expression, she'd follow with, "Sorry." No more than a minute later, though, she'd add, "and my boobs hurt so bad when my milk came in, I thought they were gonna bust out!"

I once had to pull her aside. "Georgia, please stop scaring the other girls. They'll deliver soon enough, and maybe without any trouble at all. Don't worry them ahead of time."

"Sure, Mrs Airhart," she'd say, but it wasn't long before she described the intricacies of her epidural anesthesia, delivered too late to give her the relief she wanted. To her credit, she'd shrug and smile at my look and quiet down … for a while. The other students respected her expertise, though. She expressed their fears before they could articulate them for themselves, something I didn't adequately appreciate at the time. Instead, I was focused on how complex her life would be after her second child's birth. Georgia had a naturally sunny disposition, but her good-humored acceptance of being pregnant again while still in her teens frustrated me. I admired her optimism, yet was wary of other girls following her example.

What surprised me was how Georgia took to the shower curtain periodic table project. With her usually cheerful attitude, she completed her element blocks before most of the rest of the class. One day, after I'd reviewed the students' papers to be sure they were accurate, I asked them to place them on the chart.

Georgia was the first one there. She pulled her papers and the textbook over to the back of the room where I'd laid out the marked-up shower curtain on the floor. Sitting cross-legged on the bottom half of the curtain, full belly protruding over her feet, she turned her rectangles this way and that, puzzling over where to place them.

I pointed to one of her elements. "See the energy level number you wrote here? That's the number of the row you're looking for." She nodded, but I knew she didn't yet get it. "You can also just count left to right through the spaces until you've counted the number of electrons you've written here. That's where this one belongs." I demonstrated as I talked.

Soon Georgia grinned. "Oh! I see!" she said, and she went to work taping all her element papers to the curtain with the double-sided tape I'd provided. I helped a few other students finish up and trekked back to the floor where Georgia was. When I started explaining to another student as I had to Georgia, she jumped in.

"Look here," Georgia said, pointing to what one classmate had written on her paper. "How many electrons does carbon have?" She counted out the spaces on the chart, and my first response was to ask her to let me teach the class, but I resisted. Instead, I stood back and let Georgia help her classmates. She genuinely relished sharing what she'd learned. Letting her help me meant that I could move on to discussions of how bonds form between atoms to create compound substances based on the number of electrons in their outer energy level. Atoms crave stability and will bond with other atoms with the express purpose of completing their energy levels.

The bonds formed in chemical reactions share some similarities with the bonds our students formed. When a young girl grows up in a chaotic household with few, or indisputable, rules and more experience with confusing demands than loving gestures, she has a skewed sense of love. Bonds form as a result of attraction between positive and negative forces, as they do in compounds. When a boy—an oppositely charged particle, let's say—professed love, she might well bond with him if her energy level is incomplete. An unexpected pregnancy could result.

Our girls experienced a variety of family bonds that could sometimes be confusing as well. Some were kicked out of their homes when they became pregnant. Other families welcomed or accepted the pregnancy to differing degrees. We also had a few married students every year, but marital bonds weren't always resistant to the stresses of late adolescence and parenthood. Husbands could be loving, generous mates, even if not always involved fathers. In some cultures, coupling at young ages and bearing multiple offspring were encouraged. Mothers were the acknowledged caretakers, meal preparers, and house cleaners in those cultures as well, putting further stress on our students' academic success.

***

When our shower curtain periodic table was complete, Georgia helped me hang it on one of the long classroom walls opposite the windows. She was proud of its dramatic color and artistic effect.

"It's so pretty," she said when we stood back to view the class's handiwork.

"It is," I responded. "You all did a great job."

Campus staff meetings were held in my classroom every Wednesday morning because the added lab space in the back made mine the largest one on campus. My principal and fellow teachers oohed and aahed over our wall art, and I gave well-deserved credit to the girls for the neat and variously colored craft paper blocks they'd attached to the white shower curtain. Students in my other classes were similarly impressed. "Why didn't we do this last year?" some asked.

I made a point of referring to our periodic chart wall hanging during succeeding lessons that semester instead of the chart in the textbook. It was large enough to be viewed from across the room. We talked about which elements would bond with which others and which elements wouldn't ever form bonds together because the right kind of attraction just didn't exist between them. A couple of weeks later, I gave a unit quiz and moved on to the Kitchen Chemistry unit on chemical reactions.

In this and succeeding units, I found more useful ways to direct Georgia's desire to be admired by her classmates via interruption and spontaneous comments, which didn't diminish over the course of the year. She not only enjoyed being the center of attention but also enjoyed being an unofficial teacher's aide. At the same time, I learned to value her helpful instincts and knew that if Georgia understood the principles, she could help me ensure the others did, too. She became my ally in the classroom. With Georgia orbiting my energy level, I could move up to the next level with greater confidence and wait for her—and the rest of the class—to catch up. I learned to appreciate more organic lessons, what I would later call "just in time teaching," assessing in real time what students already knew, or didn't, and moving them forward from there. First, I had to disengage from rigid lesson plans enough to make room for flexibility. I had to let students help teach each other.

8
Divided: Take risks but know your limits

Like atomic structure, cell division was a difficult concept to teach. Anything that couldn't be seen was theoretical to my students until I could demonstrate that it existed. I found the best way to make the process real inside the classroom was to view stained microscope slides of cells in various stages of division. The best cells to study were onion root tip cells because they reproduce rapidly.

On the day of our microscope lab, I projected a picture of stained onion root tip cells on the Smart Board (interactive whiteboard) at the front of the classroom. "Your body's cells are dividing all the time, just like these onion cells," I said. "See how each of them is in a different phase of division?" I'd set up our two microscopes at the back of the room and put prepared and stained slides on them for the girls to observe. "I'd like you to find and draw on your graph paper one cell in each of the four different phases of mitosis." In a fit of optimism, I purchased a couple of sets of prepared slides with plant and animal cells the year before.

In my work as a medical technologist, I'd spent thousands of hours analyzing cells using much more sophisticated (and expensive) microscopes. I felt fortunate that our classroom had at least budget-variety lab tools, but the beauty of cell division was lost by our cumbersome and inconsistent equipment. It was frustrating.

After discussing cell division for most of the week, I added a class session on cloning and showed students pictures of Dolly, the cloned sheep. They were curious about how Dolly looked and behaved, and I tried to connect her origins to the concept of cell division. Just like a human cell, a cloned organism like Dolly contains the same genetic material as the parent, which technically makes identical twins natural clones. So far, biotechnologists haven't succeeded in cloning humans, but it's hypothetically possible. I doubt I'm the first to think it's a shame teachers can't be cloned so they have a snowball's chance at fulfilling all the expectations of their districts, their administrators, students, parents, and community. Teaching four subjects a day, including two concurrently, strained my ability to switch gears when needed. I could've made good use of a clone in my divided classroom.

"As cells grow, they arrange themselves into a matrix to form living tissue," I said to the girls peering through the binocular eyepieces of the microscopes and pointed to the Smart Board again. "They might look like they're lining up in rows and columns. Cell nuclei should be visible in the different phases we talked about."

"What's that little green thing?" Jeanie asked.

"Let me see." Jeanie pulled back so I could look through the eyepiece. "What we're looking for should be purple."

After turning the fine focus knob up and down a little, I could see that the bright green blob was in a completely different plane from the onion cells. "That's an air bubble between the slide and the cover slip." For the tenth time in as many minutes, I brought the focus back to the onion root cells. "Take a look now. There are several cells in this field in different phases of mitosis."

Jeanie pulled her long, wavy hair back with one hand and leaned on the other elbow, her unwieldy baby-belly bumping the edge of the lab table. I'd asked the girls with long hair to pull it back with elastic bands, but Jeanie had insisted she didn't have one and refused the one I tried to give her from my desk drawer. Consequently, it kept falling over her face and around the microscope.

"Don't touch the knobs," I cautioned, as Jeanie repositioned herself and jostled the table while reaching for the microscope with her free hand. "It's really easy to lose your focus."

Meeting the expectations of the district's curriculum map, which specified which concepts should be taught when, was stressful. I couldn't just ignore their maps because my performance was judged at least partially by how well I adhered to them. However, my students often lagged behind the rest of the district in moving through curricula. Students could get bogged down by disputes with their babies' fathers, a colicky infant, or decisions about how to give birth (Birthing center or hospital? Epidural? Circumcise, or no?), not to mention pregnancy leave absences. All these circumstances hindered learning. Sometimes I got caught up in satisfying the deadlines imposed on me and became frustrated by my students' comparatively slow progress.

It was easy to lose focus.

***

"I've had two students out again all week," I'd complain in our Wednesday staff meetings, because this was almost always true. In a class of only four or five, these absences could derail plans for days. "I don't want to get too far ahead of them, but that slows everyone else down. Until Rachel's baby gets over pinkeye, she won't be back either." My new student, Marla, miscarried a couple of weeks after enrolling and returned to her regular school, reducing my biology class down by one. But we were expecting a new student on Monday. The months after school breaks were the busiest. Girls could enroll as soon as they had a positive pregnancy test, and long school breaks tended to ratchet up the odds of those telltale blue lines or plus signs.

"I hear you about Rachel," the social studies teacher would say. "She's missed almost a week already. We had a major test on the post-Civil War Reconstruction era yesterday, and she'll have to make that up."

Everyone groaned. Because our students were also moms, they missed school for their babies' illnesses as well as their own. There was never a day when all students were in class.

"Madelyn's so distracted by waiting for labor that she might as well stay home," someone else would say. "She's already had two false starts."

The counselor might present an urgent plea. "Farah's dad just lost his job, and she's trying to get more hours at Taco Bueno in the afternoons when her mom can watch the baby. What does she have sixth period? Is it too late to get her into the district job credit program so she can leave early and pick up more hours?" Students received an hour of course credit for one hour of work outside of school each day if they fulfilled the district program's requirements.

The weekly meeting almost always ended with a discussion about getting students back on track academically. We sympathized with their complicated lives, but we wanted these girls to graduate. A diploma was their first defense against a life of escalating poverty and hardship. We never forgot the stakes.

"End of Instruction exams are coming up soon," the English teacher might add. "Could we have an all-school assembly on test-taking strategies?"

"Hmmm. I think there's something on the district website," someone else would answer. "What about adding another Academic Options period one day next week?" Every teacher's goal was to help students complete the state's graduation requirements; we were all invested in their success.

"How about an assembly for students on learning strategies?" I suggested more than once.

Genell Coleman, who'd replaced the former principal, JoAnn, in 2008, sat in on staff meetings, but she usually held her comments until the teachers, counselor, campus nurse, and childcare teachers finished their reports. We were all aware of her presence, however. Her refrain in any conversation about teaching strategy was the same: "Engage them bell to bell, Mrs Airhart," she'd say. "Bell to bell."

I wasn't sure Genell fully understood the distractions we had to deal with in the classroom, like breastfeeding moms (in class, sometimes), the spotty attendance, and interruptions for appointments with the counselor, nurse, or WIC office down the hall. Some girls had to stop the medications they depended on for mental health conditions such as depression, ADHD, or bipolar disorder while they were pregnant or breastfeeding. Raging hormones were bad enough, but by adding medication withdrawal to the mix, we sometimes witnessed some bizarre mood swings, including at least one suspected suicide attempt. It was frustrating to be solely responsible for capturing my students' attention and maddening that I couldn't expect them to focus on lessons because they needed the class credit to graduate. While I recognized the extraordinary distractions they dealt with, I felt the girls should take responsibility for their own learning.

Given their own preoccupations with pregnancy and parenting, I now wonder how they ever managed to think about science. Students were interested in how genes determine which parent's eye or hair color the child would inherit, and this gave me at least one entry point. I emphasized the miracle of incubating a human organism inside their bodies where cells were happily dividing and developing organs and tissue.

"They're dividing as I speak," I'd say. "Isn't that amazing?" They weren't as amazed as I was. The concept of identical twins, or natural clones, was more likely to catch their attention, however. One of our students gave birth to identical twin girls during my second or third year of teaching, and the entire staff and student body were captivated by Candi's twin daughters.

I used Candi's babies as a foundation for a discussion of twins and clones, describing how the fertilized egg divides to form twins in the earliest days of pregnancy. Then I had students debate the pros and cons of ethical issues that cloning brings up, but the lesson just plain flopped. It didn't have anywhere near the impact those two perfectly beautiful twin girls had.

***

In 2009, after I'd taught in the district for a couple of years, the science department head called to ask if I'd spend the rest of the spring semester filling in at one of the district's other campuses in the afternoons, North Intermediate High School. Everyone just called it North. I'd just finished my last class and was packing up my bag to go home.

"It's only two sections of physical science," Kristi said, "since sixth period would be your planning period. I'll take the three morning classes."

"What's the situation?"

"They've already had two full-time teachers and a long-term sub resign."

"That doesn't sound good. Why so many?"

"The teacher who started the year moved," Kristi said. "The second one only stayed a few weeks, then took another job right after semester break. And we're just having trouble keeping subs in general." I knew that much was true.

I was already overwhelmed by my four morning classes, physical science being my least favorite. "Can I think about it?"

"Sure. There's a sub there for the rest of this week." She hesitated. "If you could start Monday, though …"

"I'd like to think about it and talk with my principal. I'll let you know tomorrow." As a fairly new teacher, I was eager to prove my worth in this large suburban district where our school was separated from the main campuses. I had few opportunities to interact with the other science teachers, and I missed having colleagues to share resources and troubleshoot common problems with as I had when I worked in medical labs.

"I can't think of anyone I'd rather share this job with," she said.

I agreed it would be a pleasure to share a job with her, but I was a little surprised. We didn't know each other very well and had only interacted at department meetings. Most of them were dominated by the high school science teachers—the *real* teachers—and I couldn't remember Kristi ever speaking to me personally before. That said, she was a personable and enthusiastic teacher. Maybe I could learn something from her.

"You'll be fantastic," Kristi said, though she'd never observed me teaching. I could almost hear her smile as she said this, but I couldn't tell what kind of smile it was.

***

"I need to use the restroom," a tall, lanky boy with slightly mussed dark hair said as he approached me. His faded t-shirt was stretched at the neck and at the hem as though it had recently been worn by a sumo wrestler. He looked at the wall behind me when he said this. These were the first words to greet me in my new classroom at North.

I'd rushed over as soon as my third period ended at MHP and had barely gotten settled. I didn't know what the school policy was for restroom passes yet. I didn't even know where the restrooms were.

"Can't you use the restroom during lunch?" Their lunch period, which had just ended, was a leisurely 35 minutes. Mine consisted of half a sandwich or a granola bar in the car on the way over. The distance between campuses was less than two miles, which gave me about ten minutes to wolf something down.

"Not enough time," he mumbled. He didn't even tell me his name.

"Okay," I said, "but don't be long. We have a lot to do today." I was too distracted by the other 30 or so kids in the class, most of them staring intently at me to assess their new teacher. A few sat quietly, waiting to see what would come next, but the rest chattered noisily in groups. I launched into introducing myself to the class and forgot to notice when the young man returned.

A couple of days later, the same student, whose name I'd learned was Dylan, asked to go to the restroom. The bell had just rung, and I was still unloading my book bag, so I waved him on. He returned much later with a canned soda.

"Where'd you get that?"

"Downstairs," Dylan said with a smile. Some of the class snickered at his audacity.

I was not amused. "No more restroom visits," I said. He merely shrugged.

***

My largest class at MHP usually had no more than nine or ten students. Each of the North classes had more than 30. Students seemed convinced they could run me off as they imagined they had done with their previous teachers, thus saving themselves the unpleasantness of learning any science. I tried icebreaker activities so we could get to know each other, but it soon became clear they knew each other well enough already and had little desire to know a new teacher.

I tried responding politely to their fidgeting. "Please sit down. Please keep your hands to yourselves. Please be quiet." My requests got louder but remained ignored.

I then tried warnings. "It looks like most of you have finished your assignment already. I must not have given you enough to do. Let's do the exercise on page 127 as well." This was met with groans from some, shrugs from others, and shushes from the few who understood the consequences.

I tried calling parents and even contacted a football player's coach, at the suggestion of the principal. The coach threatened to make his player sit out a game, and the young man, Doug, became more compliant … for a few days.

I'd finally reached a respectable level of expertise in my fledgling third career, but this seemed more like crowd control than teaching. Since I was already teaching physical science, I could use some of the same teaching materials at both schools. However, there were dramatic differences. At MHP, the girls could be sullen and unproductive, but very few were downright disrespectful. I also knew the other staff and principal well enough to seek counsel or share strategies with them. I didn't really know anyone else at North.

At North, the one shared lab on our floor had to be reserved, and it was a complicated process, so I never used it. Reservation requests had to be submitted well in advance, and being granted access depended on how many other teachers requested it. That meant I couldn't count on it. I had a hard enough time knowing what I'd do the next day or two. Next week was often too far in the future for me to predict. Most of our lab work could just be done in the classroom as well if I brought in the needed supplies, and contemplating taking this pulsing mass of hormones down the hall without tripping the fire alarms gave me chills. There was no Smart Board in the classroom as I had at MHP, and I had to resort to creating transparency films to present lessons on an aging and cantankerous overhead projector. Our smaller campus was fortunate in having fewer classrooms to divide district funds by, even though the ration was much smaller than it would have been for North. At least the five MHP teachers could come to a consensus on what items or equipment could best be utilized by most of the students. During my weeks at North, my evenings and weekends were a never-ending race to keep up with the physical science teacher next door, who was at least kind enough to share some tips and friendly conversation after school.

I'm convinced Kristi offered me the position merely because I was the only teacher in the district available to fill the slot, and despite my suspicions about her motives at the time, I was gobsmacked. Students never stopped moving! Physics teaches that perpetual motion is impossible without an energy source, but I'm not so sure. Trying to teach in the few moments between the kids talking (definitely not sotto voce), shoving and poking each other, and flinging pencils up toward the ceiling tiles to see how many would stick was exhausting. I don't know where they got their energy from. Even if I hadn't been responsible for teaching science while trying to keep some semblance of control, it would have been too much for one teacher. Where was my teacher clone then? When the sixth period bell rang, I tended to sit at my desk with my eyes closed for several minutes absorbing the quiet before I could begin the daily tasks of grading papers and updating online grade books and attendance records, which were due within ten minutes of the starting bell for each class, but which I submitted late. (Every. Single. Day.) I didn't know how other teachers taught in this environment; I honestly still don't know how teachers manage classes of this size. They deserve a great deal more respect than they're offered on this point alone.

I scoured books about classroom management and consulted my peers at MHP for advice. One loaned me a book about setting a firm but friendly tone on the first day of school, but it seemed geared toward elementary teachers. The chaos was ingrained at that point anyway. I even wore a skirt and heels once because I read that the clop-clop-clop of heels on tile floors and professional dress would lend me an air of authority. All I got were throbbing feet to distract me from my throbbing head.

The administrator who imposed discipline penalties wasn't helpful either. I once recommended a student for discipline after he'd let loose with several choice obscenities. However, when I sent him to the administrator's office, the student simply sat there for the rest of the period and wasn't punished, other than missing out on a spellbinding lecture on the anatomy of a sound wave. This was the same student to whom I'd assigned an F for his report on simple machines. There were whole passages that matched word-for-word from another student's report. I failed both students' papers with a brief note to come and see me. The girl didn't bother, but Stephen was incensed by my presumption that either he or Tanya had copied from the other.

"I hardly know Tanya," he complained. "I didn't copy from her."

"I'm sorry," I told him and pointed out a few identical paragraphs that I'd highlighted on both papers. "Whether you copied or she copied, you'll both receive zeroes." A few days later, when entering a sentence or two into Google, I realized both students had copied their reports verbatim from Wikipedia.

***

Within weeks, I recognized I'd never develop the same rapport with the North students that I had with my girls at MHP, which could sometimes be more like mentoring. I erected an impermeable barrier to divide the two halves of my days. The settings were so different; it felt like I was juggling two day jobs and failing at both. This doesn't count the evening English class I had begun teaching at the local community college to adults who were engaged and receptive. I began to more deeply appreciate my adult and teenage pregnant students.

When North students learned where I taught in the mornings, they were curious. Everyone in the district knew that's where girls went when they turned up pregnant. Some called it the "pregnant school." I was peppered with the questions.

"Hey, Ms Airhart, has anyone gone into labor in your class?" one girl asked.

"No," I said, without turning from the drawing on the board showing a tennis ball bouncing off the floor. "Now, when does this ball exhibit potential energy, and when is it using kinetic energy?" I traced the up-and-down path of the ball with chalk.

"Did anybody's water ever break at school?" Several girls giggled. "Ooh, gross!" someone squeaked.

I put down the chalk at this point, dusted my hands, and turned toward the class. "Let me make this clear," I said. "I don't talk about you to my students at MHP, and I won't talk about them here." This might have been the most effective message I delivered that semester. There were no more questions.

***

I only taught about a dozen weeks at North, but it seemed an eternity. I'd mostly agreed to take the assignment there to prove competence to myself and my science department chair, and I'm not sure I succeeded. As an experiment in teaching within a standard, high-density science classroom, it was an utter failure. It became instead a painful lesson in accepting the limits of my capability. That knowledge was instructive, as every failure in life should be. I'd already stretched those limits enormously when I chose to teach science later in life to teen moms. While I still had a lot to learn, I'd accomplished more than I could have imagined. More importantly, I developed a stronger appreciation for the unorthodox classroom setting I'd become accustomed to. I was content I was where I needed to be and returned to MHP the following year with a renewed sense of satisfaction in where I was placed. I'd learned my passion for novel experiences, and my ability to divide my time between different pursuits had limits.

9
Skeletal system: Schools provide valuable support systems for students

I did everything I could to ensure my Anatomy and Physiology (A&P) students at MHP had a reasonable appreciation of the 206 bones making up their skeleton before they left my class. Thank goodness for bones and joints! Without a skeletal system, we'd all be creeping blobs of skin, blood, and guts—not a pretty picture. Our bones and the joints that connect them provide places for our muscles to attach, connections that allow for more efficient movement than a paramecium. Bones store minerals like calcium and phosphorus for use as needed because they're essential to body function. They harbor blood cell production factories within their marrow for delivering oxygen or fighting disease and serve as repositories of the stem cells that determine your unique DNA. Another vitally important function of our skeletal system is protection and support.

The rib cage and sternum protect vital organs like the heart and lungs, which nobody can function without. Modern medicine has devised dependable methods of sustaining life without fully functioning kidneys or GI tract, but heart and lungs? Not so easy. And imagine how vulnerable the heart would be if it were merely covered in layers of fat and flesh.

MHP provided a different—but just as vital—type of support and protection for our students and their children. We supported girls in their decisions to give birth to the babies they'd conceived. We supported their decisions to breastfeed or not. Our on-site nurse actively encouraged it, and teachers allowed students to breast-feed in class, despite the distractions it caused. We supported their evolution from ordinary teenage girls—whose main concerns had been self-image, rocky relationships, and stubborn, if not sometimes dangerous, efforts to gain adult independence—into competent young mothers. We protected them from the harass-ment they encountered at the high schools they'd come from and protected their children via top-notch certified childcare, with a small adult-to-child ratio and innovative learning programs. While some students left us with the same challenges they'd come to us with, for a year or two we demonstrated behaviors that were sometimes new to them: respect toward each other, a healthy lifestyle, and genuine compassion. In addition, a counselor and childcare staff mentored girls in achieving wholesome relation-ships. Our Parents as Teachers and Focus on Fathers and Families programs encouraged responsible parenting behaviors.

On Fridays, Kesha, the full-time counselor, stationed herself near the main doors to bid students and babies farewell before the weekend. "Make good choices," was her parting phrase as each student left. They were so used to her exhortation that many beat her to the punch. "I know. I know. I'll make good choices." If Kesha eyed them with suspicion, as she sometimes did, they'd say with a stricken look. "What? I will. I promise." They weren't always convincing, but she did the best she could.

Most students responded positively to the protection we offered, whether they understood the strength of the shield or not. Like a rib cage and sternum, while they were in our care, we shielded them from the insults so common among their peers. Even worse, we sometimes needed to shelter them from their families. We tried to protect their hearts. Most appreciated what our "skeletal" support provided them, even if it was hidden from view.

Near the end of the first semester, just before the final test on the skeletal system, I challenged my students to assemble a synthetic, anatomically correct adult human skeleton I named BOB. BOB was my personal acronym for Bucket O' Bones, which is how this disassembled skeleton was stored. Before this point, we studied textbook content, played digital activities on the Smart Board by placing bones in correct positions, and named major bones on diagrams. Students completed anatomy coloring pages and lots of printed exercises. The skeletal system chapter was the first of the major organ systems we studied, because it was relatively easy for students to understand. We could feel our own bones, see evidence of them under the skin, and view bone marrow cavities in the bones of the sirloin steaks we consumed. Bones are concrete. My A&P students were juniors and seniors taking the one optional science credit on top of the two specified science courses required to graduate. They handled abstract concepts better than the younger students I taught. In general, though, concrete topics were easier to teach.

The morning of our BOB activity, I'd pull two or three tables together and dump the life-size plastic bones across the tops. Sometimes I warned students about the activity the day before but not always. I had a slightly wicked streak and delighted in surprising them from time to time. Breaking up the class routine could be entertaining and fed my need for novelty.

It usually took the entire class period for students to connect the bones by interpreting illustrations in the text and associating them with the tangible, touchable bones on the table. Much of anatomy didn't allow that kind of connection. Students understood the functions of the other systems—digestive or cardiovascular, for instance—but they couldn't see the organs in the same way that bones could be held and manipulated.

I also used a fully assembled desktop skeleton model, suspended from a metal rod in such a way that I could rotate it to demonstrate body sections and bone positions during lectures. Students named the skeleton "Gerald," after their former beloved math teacher who retired after my first year at MHP. I'm not sure Gerald would have been pleased, but he was gone by that time and never knew of his dubious honor.

In addition to the two skeleton models, we had a life-sized human torso model in the classroom. The torso was cut to reveal removable major organs and a few bones. We nicknamed the torso "Jackie." It came with interchangeable female and male reproductive organs; unused organs were stored in a drawer underneath the torso. The girls expressed their wicked desire to surprise me by changing them out from time to time to see if I noticed. I'd call on Jackie occasionally if the class didn't know the answers to a question in class, and he or she—depending on the gendered organs on display—could usually supply it.

On the day the girls assembled BOB, I'd sometimes find a video of someone singing "Dem Bones" or "Dry Bones" to play in the background. I steered clear of the Alice in Chains version. While my students would likely have been thrilled—and greatly distracted—by it, it's a little too grunge-rock for my taste. There are plenty of tamer versions of the song online for younger audiences. And let's face it, the traditional tune is catchy.

When a student picked up a femur, I'd sing (laughably off-key), "The thigh bone's connected to which bone?" After she placed it correctly at the acetabulum of the pelvic bone, I'd follow up with, "And what do you call those bones?" Or I'd point to one of the prominences on the femur and ask, "What major muscle attaches to the greater trochanter?" Non-threatening, low-stakes questions and open textbooks created a more relaxed environment that girls seemed to enjoy as much as I did.

A photo of BOB, splayed out on the tabletop and perfectly arranged on my classroom tables, made an issue of the yearbook one year. Students were proud to remember the names and shapes of bones. While they needed time, they required minimal prodding to complete their task. After they finished putting BOB together, they were given one more objective: determine whether the skeleton belonged to a male or a female. After measuring the angle of the pubic arch, students correctly concluded every year, as a forensic scientist would, that due to its wider angle, this feature of BOB meant that "he" was likely female.

***

Providing a supportive environment for these students, set apart from their high school peers who had different priorities and interests, gave the girls a chance to develop as both conscientious young adults and responsible moms. I would've been thrilled to teach forensic science, pure chemistry, or genetics courses. However, I recognized that more important for our students' futures were the courses required for graduation, plus practical courses like child development, office administration, nutrition, and personal finance. These would serve them well in both the short and long term only if they'd acknowledge it. I came slowly to the recognition that honoring the girls' priorities was important too. I often said that I had higher hopes for my students than they had, but I had to admit in the end that their goals were their own. Achieving them was a success, whether I valued the goals or not. Likewise, success was success whether it included a love of science or not.

A separate campus and small class size also provided protection from the assaults of teenage viciousness and bullying. Our students weren't angels, but with so few of them, there were fewer relationship dynamics to deal with. Staff was serious about each girl's positive experience, and we worked together to provide a safety net when one got thrown out of her home, got dumped by a boyfriend (even worse was when one discovered their baby's father had fathered a child with another girl), or discovered they were pregnant a second time. Every possibility was realized in at least one student each year.

Our staff was deeply invested in our students' well-being, and that investment took an emotional toll on each of us. Our best support system was each other. I looked forward to weekly staff meetings as much for the encouragement my colleagues offered me as for the academic and emotional support we provided to students collectively. Personal attention, a more forgiving attitude toward attendance and late assignments, and a host of second chances provided space for girls to succeed. The skeletal structure our campus created helped students store away vital information for use as needed like bones store nutrients and blood cells for future use. The healthy connections they formed with peers and staff often became permanent and allowed them freedom of movement both in society and their careers with greater confidence. This strong foundation, in turn, made it possible for students to support and protect their own children and approach motherhood with healthier attitudes.

10
Life cycles: Prepare students to move independently into the world

"See how pretty? Look at that!" Celia said, holding her son in her arms and pointing to the net enclosure. The boy reached out to poke at the cage, fascinated by the fluttering inside. The kids and their moms formed a circle around me, and I moved around the circle so everyone could see the Painted Ladies. Even though this was a project of my biology class, the whole school was invited.

Moms held their children's hands or perched babies on hips. We'd taken a short break from an ordinary school day, something everyone appreciated. Babies chattered, excited to see their moms mid-morning, and amazed by the beautiful, winged creatures in our cage. Some jabbed fingers at the enclosure, and some shied away from it—and me—when I passed by.

"Butterflies," I said. "See the pretty colors?" The orange, black, and white insects fluttered up against the netting, testing their boundaries.

We'd gathered in the fenced playground just outside my classroom window where the children tested their own boundaries with the playscape installed there. Teachers from the toddler classroom brought the children out to help us release our crop of Painted Ladies. There's nothing like the excitement of a child over a small thing, especially if that small thing is a miracle.

Our campus was unique from all others in the district in most respects. Not only were all our students either pregnant or parenting, but babies were everywhere. They were in the main hall before classes started. Babies were in classrooms if Mom was breastfeeding. They joined their moms at lunch time, when the cafeteria grew loud and lively with the sounds of moms and babies eating, playing, and laughing together over the meal brought in and served by the district's Nutrition Department each day. Girls pulled high chairs and their own plastic chairs around large round tables in friend groups that changed from day to day or week to week. Children thronged in the infant, toddler, and pre-school rooms of the childcare department and in common play areas. We could see and hear them from my classroom windows as they slid down slides, giggled at being pushed on the swings, and "putt-putt-putted" noisily around the playground on tricycles. Having their children nearby was essential to our students' well-being and brought joy to everyone on campus.

We integrated babies into lessons or programs whenever possible. The Family and Consumer Science teacher, Holly, taught a unit on child development that required each student to assist in the daycare center. For some, it led to a certificate as a childcare worker and eventual employment. The public library hosted story hours for moms and babies, and another group provided parenting and grandparenting programs for our students' families. One of my favorite biology units was about life cycles. My students created their own biomes one year with pop bottles, one atop the other, with unchlorinated water in one and soil in the other. Their mini ecosystems included fish and plants, which survived well enough for several weeks though they eventually met their demise. The life cycles of the living creatures were evident in those few weeks, but it was a bit depressing to watch our organisms die one by one.

The following year, I saw an ad for Painted Lady larvae and ordered a kit with a butterfly cage and larvae in a cup that held enough food to sustain them through their first life cycle transformation. We set up the cage, a round net stretched over a spiral-shaped wire frame with a clear plastic, zippered top. And we waited.

One of the reasons this project appealed to us was the inherent promise of transformation. The promise of transformation was something our girls anticipated with trepidation. MHP accepted students of any age or grade from several area school districts as soon as they had a positive pregnancy test result. They—and their babies—stayed for up to two years, although an exception could be made if a student was in danger of dropping out or failing to graduate.

When a girl first enrolled in MHP, she was often frightened and defensive. Although each girl was unique, having her own needs and talents, she'd been living within a relatively normal American teenage spectrum until she learned she was pregnant. She was transformed, within the few minutes required for one sperm and one egg to unite, from simply a teenage girl to a pregnant young woman, a mother-to-be. For months, she didn't look much different, and some girls denied reality this long and longer. No one could know, including the girl herself, how her life would change. Only teen moms were allowed in the program. If a miscarriage occurred, the student was required to withdraw; this only happened once in my memory. When a student enrolled, she'd already made the choice to carry her child to term. In some cases, her parents or lack of access to abortion made the decision for her. Regardless of our individual opinions, our job as teachers was to support a girl through an uncertain period of life. The only thing certain was that the course of her life would be permanently altered.

***

We put our butterfly cage with the cup of caterpillars and their food—a sticky, sweet concoction—on our plant lab shelf under a grow light, a fixture of my classroom that had been installed before my arrival. I have no idea what the previous teacher used it for.

"Will the light help the caterpillars grow?" Jeanie asked.

"No. But we can see what's happening better under the light." The instruction sheet said to keep them out of direct sunlight, so we only turned the light on during class when we were examining our wriggling little caterpillar herd.

Eventually, each caterpillar made its way up to the lid of the cup, spun a thread to attach itself, and formed a chrysalis, or pupa.

Creating a chrysalis was its first transformation. Students were amazed at how small the chrysalides were.

"How can a butterfly come out of that little thing?" Sadie asked. "That's much too small," she insisted. "What about their wings?"

"Well," I said, pointing to her classmate who was near term. "How does a baby with its arms and legs and little baby booty fit inside your uterus?" All the students, those who'd delivered and those who hadn't, rubbed their bellies, contemplating how it was possible.

One day, we came to class to find a couple of butterflies had emerged. Students were disappointed to have missed this second transformation, but a few vowed to come earlier the next morning so they could "catch them in the act." Finally, we were able to observe a chrysalis split open and see crumpled wings unfurl and stretch to their full extent. The caterpillar's body was transformed into a head and thorax, a ten-part abdomen, a two-part proboscis, and six long, spindly legs. The girls were thrilled to witness the births of their butterflies.

Although butterflies don't eat the day they emerge, they need to feed on sugar, whether from flower nectar or fruit, after they've dried their wings and become active. Our instructions suggested we provide fresh fruit, so I brought in a slice of watermelon, which the girls cut into small pieces and placed in a Petri dish. For extra measure, we added a little sugar water on top of the fruit. Students were fascinated to watch the two halves of the butterfly's proboscis merge into one tube and uncurl to drink from the sugar-water-fruit-juice at the bottom of the cage. Within a couple of days, all the chrysalides were empty, and I called childcare to schedule a time the next day for the children to help release our butterflies.

***

Painted Ladies aren't large. They have a 2–3 inch wingspan, and pale orange coloring on the topsides of their wings along with black and white markings. The undersides of their wings are less colorful and mostly light gray-brown with black spots. Because of these muted colors, they're disguised when their wings are closed and less likely to draw the attention of predators. They have a wide range of habitats across the United States and feed on many nectar plants. They're not picky. Yet even a hardy species like Painted Ladies, whose scientific name is *Vanessa cardui*, requires certain care just as teen moms do. Pregnant teens are prone to becoming overweight, the onset of early labor, or depression. Some battle acid reflux or hypertension. Even the healthy ones need and deserve extra care.

***

The morning we met in the playground to release butterflies, it was overcast and humid after having rained overnight. We hoped to launch our babies before the rain came again to complicate their initial journeys into the world. Butterflies require sunlight to warm themselves and to navigate their flight. On cloudy or stormy days, they can only fly short distances to trees or other sheltering spaces and await sunlight. We all thrive better in the sunshine.

I took a load of pictures of our butterfly release with the children, photos immortalized in the yearbook that fail to capture the squeals of childish delight that rang out that day. There were looks of amazement on the children's faces and on those of their moms. Some little ones reached out to touch the fragile butterfly wings, but others held their hands firmly behind them. Both reactions are fine; we all respond differently to new situations. I set the cage down on the grass and unzipped the lid. A few children toddled forward to watch the butterflies as they fluttered toward the open space. Once the Painted Ladies realized they were free, they flew up above us as the children babbled and pointed. Before long, the cage was empty, but the mood was festive as butterflies disappeared into the open air. Soon they weren't anywhere to be seen, and the children and moms turned back to their everyday activities. Shortly after the last butterfly left the cage, the sun came out.

"Mommy!" some of the toddlers cried, clinging to their moms. Daycare staff tried to pull them back into their respective rooms, soothing them with promises of play inside. Moms kissed babies and handed them over to staff or buckled them into the six-seat buggy with the red-and-white striped canopies that staff used to take babies for a stroll. Soon the babies had all disappeared into the building, and the girls and I headed back into our own classroom, satisfied that we'd witnessed one small life cycle. It was a transformation that we had a great deal of control over and one that had resulted in a crop of healthy butterflies, released to do what butterflies do: search for food, shelter, and mates. It was not unlike the uncertain futures my students faced as they navigated school, parenthood, and whatever lay beyond.

All these years later, I'm a committed butterfly farmer. I collect Eastern Black Swallowtail caterpillars from the rue and parsley I plant for them in my butterfly garden. Observing the life cycles of these fragile yet tenacious creatures, and keeping them safe until their successful launch into the world, activates a protective instinct in me. When a butterfly emerges wounded or deformed and later dies, I grieve. When I release one into my garden of pollinator nectar and host plants, I feel a sense of satisfaction that I can sustain them at all stages of life. I've invited neighborhood children over to help release my "babies," and I'm gratified by their giggled delight when a Swallowtail or a Monarch rests on a small, outstretched hand, complete with the tiniest tickle of feet, for a few seconds before it takes off. I'm reminded of the dozens of girls I observed transforming and launching into their own wider worlds, the launches of their children into adolescence and beyond. I wish them each a joyous flight.

Word Roots

Anthropo- human
Auto - self
Bio - life
Cardio - heart
Derma - skin
Geo - earth
Hydro - water
Magni - great/big
Mono - one
Terra - earth
Thermos- heat

11
Nomenclature: Learn correct names, then use them

"The knee is blank to the foot," I said and waited for a response from my A&P students.

"Superior?"

"Right! What about this one? The lungs are blank to the elbow."

The class was stumped. Finally, Riley said, "Inferior?"

I shook my head. "Try again." We'd been at this for ten or fifteen minutes already, with a diagram of the human body projected on the Smart Board at the front of the room. It was a duplicate of the image in their textbook.

"I don't know," Ella said. "Deep?"

"Melanie?" She'd been quietly studying her textbook, a curtain of salon-dyed maroon hair shielding her eyes as it fell forward over her brow.

"Medial." It wasn't a question.

"You got it! The lungs are closer to the midline of the body." I ran my finger down the middle of the image on the whiteboard. "Closer to the midline means medial." I pointed to the approximate placement of the lungs. "Further from the midline is lateral." I pointed to an elbow.

Riley groaned. "It's like a different language."

"It kind of is," I said, then laughed. "Let's call it 'Anatomese.' I know you don't understand why you have to learn this, but correct terms are important."

I'd barked at a girl once for talking to her tablemate about the baby in her stomach. "That baby's not in your stomach, Britt! Good Lord! You didn't ingest a fetus!" I was passing them on my way to the whiteboard and just happened to catch her comment. Britt flinched as though I'd slapped her. "It's called a uterus!" I said and moved on with a shake of my head.

Britt didn't say much in class after that. I felt bad about lashing out at her, but shouldn't a girl who would soon be responsible for the care of an infant at least know the appropriate name for the organ that the baby's limbs and lungs were growing in?

Using correct names for body parts isn't the only standard of nomenclature I applied in the classroom. While I ensured students learned scientific terms for individual subject matter topics, I had high expectations for myself as well. I challenged myself to learn student and baby names. Using someone's name implies respect, in my opinion. During the 8 years I taught at MHP and the 13 years at Tulsa Community College, one of the first things I did each semester was create a seating chart so I could fill in student names. Outside of class, I often practiced until I could picture each seat and the student who occupied it.

"Robin," I'd say aloud, as I looked at my seating chart and forced myself to envision what Robin looked like before I moved on. Dark hair, short, thin. Maybe I'd also rehearse a distinguishing characteristic or fact about her. A daughter named Cicely, for instance.

"Marcy." Long, curly brown hair, perfectly manicured powder pink nails. Senior and married. Her son is due in November. And so on.

Using a student's name, and pronouncing it correctly, seemed an essential first element in building a relationship.

***

Beyond the semester break, when my students had moved into organ systems, they accepted that vocabulary was a necessary part of learning anatomy. Because of that first disastrous year, I'd learned to create more hands-on exercises. We labeled diagrams, colored pages in an anatomy coloring book, and made use of interactive exercises on the Smart Board. By each unit's end, most had learned to name body parts and functions. Male and female reproductive systems and a chapter on pregnancy were featured near the end of the text, and most of the girls were aware of—and fascinated by—how their bodies worked. The most com-pelling biology topic for me was the final chapter on genetics, the nuts and bolts of heredity via discussion of gene assortment and DNA replication in offspring. At this point in the year, most girls could use accurate terminology for genes and reproduction, which was very satisfying.

What students didn't need coaching on was the importance of the names they chose for their children. It was one of the most important choices they'd ever make and one that couldn't easily be changed. Students worked hard to choose unique names. They were insistent that no one else chose the same name because they wanted their children to stand out. Fights sometimes broke out if one student "stole" the name another student planned to use. Some students, on the other hand, deliberately chose common American names. These were often Asian or Latina students who chose names that were easier to pronounce. Many of them had nicknames for their foreign-sounding given names. They were eager for their children to blend in. Names were sometimes unfamiliar or foreign to me.

"Can you write it down for me?" I'd sometimes ask when a name was difficult to pronounce. In every case, my goal was to use names when referring to students.

***

"Aren't those languages dead?" one of my students asked with a hint of suspicion in the first week of the 2009 year. I'd explained how classes would begin each day. I again pronounced the five Greek or Latin word roots written on the upper left corner of the whiteboard. Each word root was followed by two vocabulary words that included the root.

"Yeah," somebody else piped up. "They should stay dead!"

I joined their laughter. "That's the problem," I said. "They just won't stay dead. We chop up their dead words and make new words out of them all the time."

Genell Coleman, our principal, suggested teachers come up with a warm-up exercise for the beginning of each class period. Linda had left to take a position at the Tulsa Technology Center by this point, and I came on full time, adding Junior English, journalism, and technology classes to my class load. As a science teacher who also taught English, I thought it fitting to analyze how our language was created and recreated from different vocabulary sources. Most of the roots and the sample words I used related to science.

Each morning after the bell rang, I went over the five weekly roots on the board, mostly prefixes and suffixes, and gave students two examples of words we use today created from these ancient words. Sometimes I had students read the words, then we discussed their meanings. Throughout the week, I made a point of demonstrating word roots in everyday lessons to reinforce the concept of using word stems to create new words. On Fridays, after reviewing the words one last time, the girls took a ten-point vocabulary quiz. Most got 100 percent correct, so it was an easy A, which meant there were few complaints.

"Morph means shape or form," I'd say. "Do you remember the prefix 'a' we had a couple of months ago? It means not or without." A few were kind enough to nod their heads. "When you put these two roots together, we get a word that means without shape or without form. That's how we get this vocabulary word, 'amorphous.'"

Unlike other challenges to student comprehension skills that sometimes fell flat, I didn't fret much over their responses and would simply reword the definition or provide a different example. Words—even apart from their meanings—bring me joy. I love breaking them apart. It brings me the same joy that slicing through a cow's heart in an A&P circulatory system unit satisfies my curiosity about how it works. Dissecting cow hearts and word roots appeals to my analytical brain. While the girls weren't all that fascinated with ancient language stems, they looked forward with glee to cutting up the cow heart I purchased from a downtown butcher shop. Students brandished those shiny new and extra-sharp scalpels with surprising confidence. This should have worried me a little, but I learned to lean into whatever excited them about the subject matter. The dissection lab became a reward for surviving lessons about blood vessels, oxygenation, and heart chambers.

On the other hand, the daily word roots exercise was a reward for me. Only me. Starting a class this way grounded me. The girls tolerated it and even counted on it as a consistent part of the class structure. If I forgot one day to call their attention to the weekly vocabulary, as I did a few times, they reminded me. Words that fascinated me or taught me something new were more likely to make my weekly list, as were words I knew we'd use in a science lesson during the week. A teacher has to take joy where she can.

As in autotroph. Self-feeder. An organism capable of self-nourishment. Teaching teenagers could sap my energy. Sometimes I just had to feed myself.

12

DNA extraction lab: Find what works and repeat as necessary

"I'm doing a lab with my students tomorrow," I explained to the liquor store clerk who rang up my pint of Everclear. With tax, it was about six bucks.

She eyed me over the top of her glasses. "Oh, really?" said the liquor pusher to the presumed alcoholic. Everclear to an alcoholic is like a mainline of heroin to a drug addict.

I couldn't tell if she was genuinely interested or suspicious, but something propelled me to explain. "We're extracting DNA from strawberries, and the lab calls for 100 percent ethanol. At 95 percent, Everclear's close and a lot cheaper." As soon as a bottle of scientific-grade ethanol is opened, it begins to attract water from the air around it, and I figured it would be degraded after I'd used it a couple of times anyway. I had a hunch Everclear would work. Laboratory-grade ethanol from Fisher Scientific, on the other hand, was around a hundred bucks for roughly the same quantity and would require overcoming some district red tape, unless

Materials

Everclear
Coffee filters
Strawberries
Baggies
Wooden Skewers
Plastic Cups

I ordered and paid for it myself. "After we extract DNA from the strawberries, we'll extract it from our own cells."

"Really?" This time, I think the clerk was convinced I was telling the truth. "Huh. That's interesting."

"It is," I said. "It's one of the only labs I know my students will be impressed by." In fact, with the last bit of ethanol in my supply closet the previous year, I'd been overjoyed to find a lab procedure that worked as it was supposed to and one that amazed the girls. I suspected the ethanol left by my predecessor had been watered down by frequent openings over time anyway, so it was likely much less than 95 percent by the time I arrived at MHP.

I took my pint bottle in its brown paper bag, top twisted around the neck, and brought it home to store in my freezer overnight. Our experiment called for ice-cold ethanol, which doesn't freeze at the same temperature as water.

The next morning, getting the Everclear into the freezer in Holly's Family and Consumer Science classroom was easy. Her room was just down the hall from mine. I didn't tell her what I'd put in her freezer. She was at the front desk when I snuck down and was in such deep conversation with the school secretary that she probably hadn't even noticed I'd gone into her room. With only four classrooms, each teacher stored some measure of communal supplies, so we often had to retrieve something from one room or another. Her classroom, where she taught cooking and other household skills, had the only freezer on campus with enough space to stash the Everclear. Plus, I could retrieve it later without passing by the secretary's desk on my way to the staff breakroom freezer and likely avoid notice.

I didn't worry Holly would disapprove of having Everclear in her classroom freezer if she understood the reason, but I wasn't altogether comfortable with explaining myself. Our principal, Mrs Coleman, was straitlaced. Because of our school district's thinly veiled derision of our student body as "disgraced," she insisted that students and staff adhere to rules and district policy. I respected her desire to keep students and staff on the straight and narrow, and recognized that she might view public judgment of our campus by outsiders as an indication of poor supervision on her part as our principal. The "Use of Alcohol, Drugs, and Controlled Substances" section of the employee handbook clearly forbade any employee from possessing alcohol in school buildings. If I was crossing a line by smuggling in a pint of nearly pure ethanol, I didn't want Holly to be implicated.

The first bell rang at 8 a.m., and I arrived at least a half hour early each day to prepare for class. As an early riser, I didn't find this difficult. On the day of the DNA lab, I laid out all the lab supplies we'd need: test tubes and racks, funnels and filter paper, wooden skewers, and the quart of strawberries. Everything but the Everclear.

At MHP, my rolling lab table was situated at the back of the classroom in a niche that included a fume hood for pouring or mixing flammable liquids and an eyewash station in case of accidents. There was a spacious storage closet and two sinks along the wall adjacent to the lab nook. What happened in the science lab at MHP stayed at MHP, for the most part. The district's Science Instructional Specialist usually visited my classroom at the beginning of the school year, but otherwise communicated with me via email from her office on the main campus—an office I never saw.

I was on my own when purchasing disposable lab supplies and a lot more. While this was costly, it also meant less oversight. I imagine at the nearby Broken Arrow High School with more than 4,000 students, science teachers could requisition supplies from the science storeroom for their labs. There was probably a form. It probably required a week's notice. You probably wouldn't know until you got into the lab exactly what and how many materials you had. I had greater control over my own lab.

Among hundreds of other materials over eight years, I bought the aquarium that housed the gerbils we installed in the classroom one fall from PetSmart, along with the activity wheel, plastic tunnels, and live crickets they consumed every week until their deaths several months later of some mysterious respiratory illness. The aquarium was reinvented as the faux ecosystem for demonstrating convection currents in an environmental science lab after we lost our gerbils, which the girls had named "Pancho" and "Tequila."

I provided lemons from the grocery store for the lab where the students made batteries to light a tiny light bulb, purchased at the hardware store. Each fall, I bought the one-per-student big bottles of Elmer's glue (clear so we could add food coloring, which I also purchased) and the 4 pound box of Borax (the remains of which are in my home utility room closet above the washing machine to this day). These were the essential ingredients for making slime, otherwise known to students as "fart putty." I hit the start of school sales at Walmart and Office Depot, coupons in hand, with the frantic energy of a knick-knack hoarder at an estate sale. As the year wore on, I estimate I spent more than a hundred dollars a month on lab supplies. Every weekend, while planning lessons and labs, I created a shopping list for my trips to the hardware store, the grocery store, or, in this case, the liquor store.

For the DNA extraction lab, I brought filter paper (coffee filters), strawberries, and Everclear. I'd considered ordering the ethanol from Fisher Scientific more than once—I still had a personal account from my hospital laboratory days—but it was too expensive. The liquor store was on my way home from school, and Everclear was cheap. When I was a young medical technologist at Lake Charles Memorial Hospital in Louisiana during the 1970s, our full-time human dishwasher gave all glassware a final rinse in Everclear to remove any lingering soap or oily film that hadn't been eliminated in the prior rounds of sterilization and handwashing. I considered the product laboratory-grade.

***

As I'd told the liquor store clerk, the DNA lab in biology class was by far my favorite lab. It worked every time. No matter how many or which students were in the class, every one of them was astonished when they could see and touch their own DNA. It was also a very simple procedure and cost less than some of the more elaborate labs I regularly funded.

The morning of the Strawberry DNA Lab, I was especially excited for class to begin. Pulling pregnant or parenting girls out of their own thoughts was hard, and many labs inspired nothing more than absent-minded compliance or faint interest. There were a few exceptions to this, and pulling their wooden stirring rods from over-sized test tubes containing a mush of soapy strawberries in cold alcohol to deposit a long, snot-like string of material clinging to it was more dramatic than most. Touching their own DNA imprinted a visual image they could carry with them after this point.

Just a few minutes before the first bell, our principal came into the classroom. Despite sometimes being didactic, Mrs Coleman was supportive of both students and teachers. I ordinarily didn't fear her presence in my classroom, but what were the odds she would choose to visit my first period biology class the day I had to smuggle a pint of Everclear back down the hall? I'm still not sure what prompted her visit since she'd already done the requi-site once-a-semester staff observation.

"Mrs Airhart, what are you doing first period?" she asked. When I explained about the DNA lab, her face brightened. "Oh, that sounds interesting! May I join you?"

"Of course."

She had a few early morning tasks to attend to, but she promised to be back in a few minutes. "Don't start without me."

I sneaked my contraband in its brown paper bag from down the hall onto the lab table just before the bell. Take the bottle out of the bag? Leave it in? I was torn. I was confident Mrs Coleman was a teetotaler, but she might recognize what the brown paper bag signified. I was even more confident that my students would understand. I decided to discard the paper bag.

I'd just introduced the lab to my students, and we'd gathered around the lab table when Genell reappeared. "Mrs Coleman wants to see our lab today too," I told them before giving them instructions for smashing their strawberries and mixing soapy liquid for extracting the genetic material from the mush of cells.

"Oh, cool!" was the standard response to the string of strawberry DNA. Strawberry cells have more copies of DNA per cell than human cells do, so there's a sizable mass of DNA in a smallish sample.

"Now, each of you take a wooden tongue depressor," I said, handing one to each of the girls, "and scrape the inside of your cheek for several seconds. When you've finished that, take a big sip of water and swish it around in your mouth for at least 30 seconds before spitting into a clean test tube." We repeated the steps we'd followed previously, adding extraction fluid to the tube to mix well with the spit and finally layering cold Everclear over the top. The cloudy layer between the emulsion and the alcohol is DNA extracted from living cells. There's not as much in a cheek swab, but it's visible, nonetheless, and recognizable as DNA after seeing what was extracted from the strawberries.

Even Mrs Coleman was impressed. "That's fascinating!" She was so enthusiastic over what she'd observed that she mentioned it to other staff members who cracked a few jokes about it at lunch. "You're the teacher's pet today, Airhart," one colleague commented. I was more excited about my students' reactions than the principal's, but I was definitely pleased she'd found the lab so interesting as well.

One thing Mrs Coleman never mentioned, to me or anyone else, was the fact that contrary to rigid district rules—rules she rarely thwarted—one of her teachers had possessed a pint of liquor on campus.

13

Dissecting food chains: Preview all class materials thoroughly before presenting them to students

"What do you have there?" I asked, leaning over Tiffany as she filled in the wings of what appeared to be a dragonfly with a gray pencil.

"Those are dragonflies, Mrs Airhart. Can't you tell?" She kept coloring.

"I can definitely tell they're dragonflies, and they're very nice dragonflies. So are those sunflowers at the bottom? You know how jealous I am that you can draw flowers that look like flowers. What I meant was what level of the food chain does that dragonfly represent, and what will you draw on the next level?"

Tiffany laughed. "You're right. Your flowers would look like cat faces or something." Her assessment of my nonexistent artistic

talent was accurate. The diagrams I'd drawn on the whiteboard to explain this exercise looked more like a murder scene gone horribly wrong than a simple explanation of predator and prey in a food web.

"These dragonflies are my first level of consumers. I'm gonna add snakes or frogs after this; I can't decide. Maybe an owl at the top."

"Perfect," I said and moved on.

Owls tend to perch at the top of their geographical food chains. In most ecosystems, adult owls have no natural predators, but their eggs and young can be targeted by large mammals or snakes. Owlets are easy marks because they lack flight feathers and other features that characterize the adult raptors they will become. Depending on their habitat, adult owls might feed on insects, fish, reptiles, turtles, and small mammals like mice or rabbits. They're fierce predators, but they can't digest all of what they eat. Consequently, they vomit the bones, hair, and other indigestible bits of their prey into a neat package called an "owl pellet."

In my biology class's study of food chains and food webs, we spent a week or more on the discussion of the "who eats who" of life forms. Our text illustrated levels of various food chains and described the amount of energy that passes from one level to the next as one organism consumes another. The energy "producers" are always at the bottom of a food chain, and all other levels are "consumers."

One graphic exercise I had students complete while teaching this topic was a pyramid-style illustration of food chains, like the one my student drew with a dragonfly, a frog, and an owl above the sunflowers. Students constructed the pyramids out of card stock that was cut and folded to produce a pyramid. I instructed them to sketch or draw a different food chain of at least three trophic (feeding) levels on each of the three sides of their pyramid. They should also note what percent of the energy from the tier below each level, on average, was transferred to the next: 100 percent at level one, 10 percent at level two, and so on. Most of the students' pyramids contained an owl.

***

In my fifth or sixth year of teaching biology, I decided to order owl pellets collected from Barn Owls during the unit on energy transfer in food chains. I hadn't ever done an owl pellet dissection before, but I'd read the lab procedure in our textbook every year somewhat longingly. Curiosity drove me to find new ways to demonstrate topics each year. Some tried and true labs became part of my permanent repertoire, but I spent an inordinate amount of time every semester exploring new, more interesting methods to teach a concept. I was eager to see what owl pellets contained. This particular year, I had a larger class than normal. It seemed worth the personal expense and effort to order supplies for six or seven students instead of only three or four, which was a more typical class. I took the plunge.

The owl pellets arrived well before the day of the lab, and I began searching for an appropriate video to show students about what they might find when they dissected their pellets. Our campus had been one of the first in the district to install Smart Boards, probably because we only had four classrooms. It was an inexpensive way to try out the new technology before installing thousands of them throughout the district. I fell in love with mine and made good use of the novel presentation methods it allowed. It was perfect for showing videos to introduce new concepts.

Perennially pressed for time when lesson planning, I did an extensive search of YouTube for applicable videos. I found quite a few that mentioned owls, but none that discussed owl pellets. Some focused on food chains but didn't address owls. Some were just too long. Finally, I found a four-minute video that seemed to both address an owl's designation as a fierce predator and briefly describe its processes of food ingestion, digestion, and ejection. It was called "True Facts About the Owl." Despite some silly jokes early in the video about an owlet looking like a Muppet or a cotton ball that grew feet, and a very odd narrator's voice, the video did indeed provide true facts. After viewing the first couple of minutes, in which an owl's dietary habits were mentioned, I copied the link into my lesson plan, and I was done.

In the morning I planned to show the "True Facts" video, shortly after class started, I got a message from Mrs Coleman that she wanted to speak to me in her office. "It'll only take a couple of minutes," she said. We weren't allowed to leave classrooms unattended, but Kristen, the English teacher across the hall, didn't have a class that hour. I asked if she'd sit in for me while the short video played. "I'll be back before it ends," I told her. I hit "play" and left.

When I got back to class a few minutes later than expected, the students and the ersatz sub were roaring in laughter. Kristen met me at the door. "They talked me into showing it again," she said, hardly able to get her words out because she was laughing so hard.

After watching for a minute or two, past the point I'd watched over the weekend, I realized my mistake. While the narrator, whose name I later learned was Ze Frank, presents facts and does describe the feeding cycle of an owl, he also comments that it's perfectly polite to throw up at an owl dinner party. Funny, but innocuous. The humor is deliberate, irreverent, and at times a bit inappropriate. Frank has produced an impressive volume of work since his 2013 owl video and has starred in TED talks and Friskies "Dear Kitten" commercials. All I knew at the time was embarrassment that I'd neglected to watch the entire video before choosing to show it to the class.

Since I failed to see that last half of the video, I missed the part where a fairy-tale girl asks an owl where her mother is, and is told, "How the hell should I know?" She then gets her face ripped off and her eyeballs eaten by the owl. I missed Frank's parting message not to do drugs "so an owl won't swoop down and rip your face off." I'm not so sure the message about what owl pellets contain was delivered in the most suitable manner.

This became a great lesson in learning to laugh at myself—or perhaps with myself. I admit the video was funny even if the joke was partly on me. The biology class insisted I show the video during Academic Options (our version of study hall) that week to other students, and I complied—once they'd finished their work. I also admit to showing a couple more "True Facts" videos in the weeks after this faux pas as student rewards for finishing assignments early. Most were hilarious.

We completed our owl pellet lab the following day, and it was much less spectacular than promised by our textbook authors. Some of the rodent bone fragments just looked like large pieces of sand or small pebbles. Nothing was distinguishable as a bone. Some pellets contained more fur than bone. I won't say the lab expense was wasted, because really, I would never have discovered "True Facts" if I hadn't bought the pellets. And this one stands out as a great find.

The owl pellet episode is burned in my memory as an unfortunate choice that provided an unexpected gift. I might have committed a teeny lapse in judgment, but I did amuse us all for a while, which is quite a feat when dealing with hormonal teenagers who can be fiercer than predatory owls sometimes. An important professional point is that I'd encouraged students to dig into the owl pellets and thoroughly analyze what they found, although I'd been remiss in fully dissecting the video used to introduce the concept. I'm over my embarrassment. For a few days, I was the most popular teacher on campus, and I admit to the ego boost that provided. Not a total failure, but still.

Advice about observing a video in full while lesson planning was probably covered in that first year of the bachelor's program in Secondary Education that I missed by becoming alternatively certified. Sometimes experiential lessons are the best ones, though. For anyone who also skipped that particular class or shrugged the advice off as irrelevant, let me suggest, as one who now knows: Preview a video all the way to the end before showing it to students.

14

Magic city: Lessons shaped by passions or convictions often have a greater emotional impact

"When you see the F-word, say 'freak' or 'freaking.'"

Students nodded and flipped through their copies of the books I'd just passed out. The year was 2012, and I was full time, having added English and journalism courses to my teaching schedule. While the classroom shelves were half-filled with books before I was hired, it's likely I bought this set myself. But I'm not certain. That's how little discretionary funding the district provided. Classroom supplies for our campus didn't weigh heavily on administrative budgets.

"Use 'shoot' for S-H-blank-blank. You get the picture." More nods.

At least a couple of students were still pregnant, bulging bellies pressed up against their desks while they rocked back and forth or bounced up and down on large pink and blue exercise balls. I'd purchased the balance balls with a bit of windfall money to help relieve back and hip pain for girls struggling with the excess weight of late pregnancy. I'd hoped it would also provide a crutch for kinesthetic learners. I'd been one myself.

Junior English, the period just before lunch, felt like the longest of the day. Pregnant students were starving and grumpy. The parenting girls were eager to pick up their children from childcare at the other end of the building so they could share lunch and conversation with them and with friends in the cafeteria. Moms who were breastfeeding had swollen, throbbing breasts. I was ready to sit down for a few minutes to rest my aching feet.

Invariably, in the middle of learning segments, we were interrupted by someone next door complaining about being too hot or too cold. The girls and I took turns increasing and decreasing the temperature on the thermostat in our room that controlled both classrooms, which meant that expecting extended periods of focus was a delusion. A campus made up entirely of women between their teens and their 60s sometimes spawned discord. Add in the 20 or so babies, toddlers, and their caregivers, and chaos could erupt. One such drama broke out in my classroom when one of the moms threatened another during environmental science class.

"I'll beat your scrawny ass!" the aggrieved student screamed at her classmate across the room. "Your baby bit mine. You can still see the teeth marks."

The accused mom jumped up out of her seat and sputtered, "You just try!"

I placed myself between their line of sight. "Calm down," I said to both of them. "Babies bite all the time. Maybe she's just teething." All students took child development units in the Family and

Consumer Science class, and most participated in one or both of the extracurricular parenting programs we offered. I was sure teething and biting were topics they'd discussed.

"She needs to teach her baby that's not okay. Bite her back when she does that."

"Hold on," I said. "Nobody needs to bite anybody back. That doesn't accomplish anything." Eventually, they sat down again, but they continued to fume. I heard there was a scuffle in the hall after class, and both students were ushered to Mrs Coleman's office. By the next day, the incident had blown over, but the student whose baby was bitten never returned to class. Patience was sometimes the only thing that kept us from killing each other.

***

I'd assigned the book *Magic City* by Jewell Parker Rhodes to my Junior English class. I wanted to explain how hard I'd had to lobby our principal to teach this book, a text I considered much more relevant to them than what the district required, but I didn't.

At issue was the obscene language, more so than the obscene events described in the book—a historic race riot and resulting massacre—which I found ironic. My students were well-versed in language obscenities, but sanitizing certain words was necessary if we read passages aloud in class as I planned to do. Mrs Coleman stipulated the rules, and I was compelled to oblige. I had to agree with her that teaching about a race riot to 16- and 17-year-olds was risky, despite its historic value.

"*The Great Gatsby* is part of the district's curriculum," Mrs Coleman said, eyeing me over her computer screen. We were in her office one afternoon just after winter break. I had waited until the halls were emptied to speak with her because I knew she'd be less distracted. "It's one of the classics," she said. "We all had to read it."

"I know," I said, "but I hate it. A story about a bunch of privileged White folks who get away with murder. These girls won't learn anything useful from that." Just because we were all forced to read it didn't seem reason enough to perpetuate the torture. And just because the book I wanted to teach included the liberal use of racial slurs and a smattering of F-bombs—in the service of exposing grave injustices—didn't seem reason enough to exclude it. *Magic City* was the violent story of a huge population of White Tulsa vigilantes getting away with mass murder against the Black community of Greenwood that thrived at its northern fringe. It was fiction, but it was based on actual events.

"*The Great Gatsby* is great literature," she said.

I shook my head dismissively. "But what does it teach? Granted, Fitzgerald's writing exemplifies period literature, but the shallow characters ... and the message ..." I grimaced.

Genell sighed. "Tell me about this book." She glanced at the back jacket of *Magic City* and scanned a few pages while I made my case.

I'd been introduced to *Magic City* at the 2004 Oklahoma Celebration of Books in Tulsa, where Jewell Parker Rhodes was a featured speaker and workshop leader. Rhodes had visited Tulsa in the late 1990s to research the 1921 Race Riot and fictionalized the horrendous events for her book, which she published in 1997.

On the morning of the May 30, 1921 Tulsa Memorial Day Parade, a Black, 19-year-old shoeshine named Dick Rowland entered an elevator at the Drexel Building operated by 17-year-old Sarah Page, who was White. The top floor of the building housed the only existing restroom facility for Blacks on that block of Main Street. When he entered the elevator, Rowland apparently tripped and grabbed Page's arm to steady himself. She screamed. Numerous opposing myths have sprung up since the event about assault, molestation, or secret assignation. The two must have been familiar with each other, considering where each worked. Page at first would only say Rowland had grabbed her arm, but she refused to press charges and there is no written record of her questioning.

It never really mattered what Page told the police. Her scream was all that counted. Tulsa was a growing city of mostly White citizens, enjoying their status as the self-proclaimed "Oil Capital of the World." Meanwhile, the Black community of Greenwood prospered at the northern edge of downtown. The successful businesses on Greenwood Avenue were known regionally as "Black Wall Street." It was the envy of both Black and White entrepreneurs. As a result, simmering tension began to boil. But by the evening of June 1, the thriving shops, hotels, and banks were destroyed. A lynching of the 19-year-old Rowland was narrowly averted. Hundreds of his neighbors were murdered, along with a few dozen Whites, and the bodies of the Black dead were dumped in mass graves. The Greenwood community had been wiped out.

Despite having lived in or near Tulsa since 1989, I'd never heard about the race riot, now more appropriately called the 1921 Race

Massacre. When I did, I was horrified. The misinterpreted inciting event. The racially motivated vigilante violence. A community destroyed. Controversial reports of low-flying airplanes, National Guardsmen, and burned-out neighborhoods. Hundreds murdered. Ten thousand homeless.

I'd completed graduate courses at the university campus just a few blocks from Greenwood Avenue and just yards from Mt Zion Baptist Church, the building that once housed Vernon African Methodist Episcopal Church and shielded many Black residents from danger during the riots. It barely survived the fires and gunfire that raged around it. I'd driven through Black Wall Street on my way to evening classes one or two nights a week for two years, completely ignorant of its history.

I was surprised that none of our students had studied this dramatic event in Oklahoma history. The fact that this level of destruction and death occurred only a dozen miles from our campus made it imperative to me that the girls know what happened there. Though we had a smattering of Asian, Hispanic, and Black students, most students were White. Almost all of them were disadvantaged. Targets of shunning and sometimes hateful comments among their peers because of their pregnancies, these teen moms were sensitive to injustices of all types and generally tolerant of the differences between them. I believed they would find the race riot as reprehensible as I did, but I had to first convince Mrs Coleman of the book's value.

"The language is definitely not appropriate," she said, closing the book and sliding it back to my side of her desk.

"I know. That's why I'm asking you first." Our principal didn't generally interfere with classroom instruction and trusted teachers to adhere to district curriculum requirements. She was dead serious, though, about the school's reputation in the district. She didn't take criticism from the administration lightly. It was bad enough that the girls suffered scorn from their peers for getting pregnant, and mainly from peers of their own sex. They reported snide looks and degrading remarks loudly whispered in their presence like, "Slut! Baby Mama! Keep your legs together, why don't you!?" There was also a district-wide ridicule of academic performance on our campus, comprised of the sometimes low-performing students that district administrators presumed to be of "questionable moral character." Our principal took those attitudes personally. District biases were unfair to the girls and unfair to our dedicated teachers, childcare workers, nursing staff, and counselors. No one bothered to ask what kinds of students we were mentoring. These were young moms with a great deal more on their virtual plates than most high school students. Following district curriculum guidelines didn't always make sense, but there could be repercussions.

"I think this is an important event in Tulsa history, and we'll do research into existing documents as we read so we can understand the context of the story." I explained the online archive of material we had access to from nearby Oklahoma State University.

Genell crossed her arms and leaned back in her chair. "Here's what you need to do," she said. "Compose a letter to your students' parents outlining your reasons for teaching this book. Make clear that it includes some inappropriate language. Let them opt their students out if they choose. If most or all opt out, teach Gatsby."

In the end, only one parent objected, and I was cleared to teach *Magic City* to the other students with stipulations about not reading the obscenities aloud. Poor Jennie was doomed to read *The Great Gatsby*, a cruel irony. I'd heard Jennie utter many of these same obscenities freely with her classmates. It may also be that none of the other students shared my letter with their parents. In any case, neither my principal nor I realized at the outset how this book would affect the entire class. I was so passionate about the book and the events it illustrated that I didn't consider what Jennie was absorbing during class discussions and lessons. Students complied with the language rules I'd set out, without fail, so at least the version of the book Jennie overheard included cleaned-up language. She'd comment on our discussions or ask questions about the story. "Wow! That's not right!"

A couple of days a week, we took turns in class reading pages and discussing character motivations, setting, plot, and themes. Some students read easily, while others stumbled.

"Entra… Entry…" Students looked to me for help with unfamiliar words.

"Entrepreneur. Someone who starts a business." She'd nod and move on. Reading aloud was a great way for me to assess their reading abilities.

Every couple of pages, I'd ask questions. "Why do you think Joe dreamed of leaving Tulsa?"

"He knew he wouldn't get a fair shake there." My students weren't dreamers in the same sense that Joe was. They mostly stayed close to home. Their dreams had more to do with a handsome boy sweeping them off their feet, but they understood discrimination. "Shoot. Tulsa hated Black people."

A few pages further, I'd ask, "In Joe's nightmare, he keeps hearing 'Who do you think you are?' How's that a premonition of what's to come?" I'd already introduced the basic facts of the 1921 events. "Why do you think the author used this device?"

"He knows the White people won't ever respect him just because he's Black."

From our classroom, we researched the online historical material available at the Oklahoma State University archives just down the turnpike from us, projecting photos and other documents on our classroom's Smart Board. Students created and shared character sketches of the book's fictional characters, in which they associated words and actions with probable feelings and motivations. In the novel, Dick Rowland's character was renamed and fictionalized as Joe Samuels. Joe was the protagonist whose perspective guided the story. My students could appreciate Joe's dilemma.

I had students compose fictional blog posts by characters to speculate on what Joe or other characters might say, and we posted them to a private classroom blog where the whole class could read them and share comments. They constructed timelines for the actual historical events and completed research worksheets, both of which provided the context for what we read. I administered short quizzes, but these scores provided a small fraction of the unit grades. I was less interested in their memorizing facts than in their understanding the cost of racial biases. I wanted them to understand a writer's power to evoke compassion for the people in a book's world—in this case, based on real people.

Curious about students' perceptions a decade later, I reached out to a few former students. "Magic City!! I couldn't remember the name of that book!" Brooke said. She continued, "From that book and research, I gathered that our education system can be a bit backward and lean towards only teaching about the good of our nation and history."

Alexis commented, "Reading difficult history passages really helped me gain new insight on how privileged and lucky I am to not have grown up in that time period." Alexis was the only non-White student in the class. "Especially being half Black, it helped me understand and appreciate my ancestors for the hard times and the struggles they endured during the 1920s."

Alexis also said of the events recounted in the book, "It's very relatable since it happened in Tulsa. I think, for me as a student, it made me want to pay better attention to the lesson." While I remember the verbal gymnastics we resorted to in order to avoid the obscene language in the book, she says she doesn't remember the language being an issue at all. "With everything kids are exposed to nowadays, a few bad words won't do any damage." Instead, she made a larger point. "It's frustrating that we still are trying to teach equality when it comes to something as simple as the color of people's skin."

Even Jennie, who was forced to endure Gatsby, learned what I wanted the whole class to learn: an awareness of historical racial injustices that led to the mass slaughter of Black citizens by White vigilantes in our own neighborhood. And that in the years that followed, the White community was almost completely silent about the event. Until the 100th anniversary, very little attention was paid to it at all.

There's still not an accurate body count of the dead. The event was renamed the 1921 Tulsa Race Massacre in most sources as the centennial observance approached, and efforts intensified to locate a purported mass grave site. Skeletal remains of 19 bodies, adults and children, were unearthed in late 2021 not far from Greenwood. They were likely victims of the massacre. Thirty-five coffins were also discovered at the same site, unmarked and forgotten by all but the Black community. Through the characterization in Rhodes' book of Greenwood and Black Wall Street, students learned about a ghastly historical event and cultivated greater empathy for targets of injustice.

A more recent flap about Critical Race Theory (CRT) has dredged up this painful episode for many because it demonstrates the unwillingness of the predominant American culture to confront the truth of history. Never mind that this four-decade-old concept, which demonstrates how the social construct of race and racial bias have become embedded in American legal systems, was never taught outside law school classrooms. Some have created a political hot potato out of misinterpreted and deliberately misleading blather about CRT being taught at the high school level, which it never has been. Most people really don't understand what CRT is, and many obfuscators prefer that they don't. Better to pretend it's a conspiracy by educators (a large majority of whom are White themselves) to make White students feel guilty for being born White. While this controversy is simply silly on one level, I think it's dangerous to deliberately hide the truth—the epitome of whitewashing—by not teaching what is arguably the most violent racial incident against African Americans in recent American history.

Ashley, the social studies teacher who taught next door to me in 2012, said she honestly didn't remember teaching about the Tulsa Race Massacre back then. With the closure of our campus, Ashley moved on to another public high school near Tulsa. "It's in the 2023 Oklahoma state social studies standards, though," she said to me last time we spoke. "I taught it last trimester and last year. My goal with all of the anti-CRT bills being passed is to stick to the standards. No one should complain about me teaching the state standards."

As long as social studies educators are in control of drafting state standards, that's a reasonable attitude. However, I fear this won't always be the case. Legislators and others are sneaking their sticky fingers into classrooms all over the country with the intent to shape curriculum to their personal views and biases, factual history be damned.

A century after the appalling events brought to life in *Magic City*, the 1921 Tulsa Race Massacre Centennial Commission succeeded in raising $18 million from mostly individual donors like me to build the Greenwood Rising Historical Center. I donated money for the placement of one brick. The Center opened in the summer of 2021 with fanfare somewhat subdued by a deadly pandemic and the disturbing timing of a former president's post-campaign rally in downtown Tulsa a couple of miles away.

The Greenwood Rising Historical Center's opening breaks the silence on this shameful event in Tulsa's history. While the wounds of Tulsa's Black citizens are far from healed, some progress has been made. I'm proud of my small part in educating a few young girls about the violence that took place in 1921, in the hopes that they and their children will embody greater empathy for people of color and all marginalized groups. There's also a brick on Greenwood Avenue today with my name on it, a small reminder of one teacher's awakening.

15
End of instruction: Honest concern for students' welfare is key to their academic success

"Mrs Airhart," Nikki said, with her usual Oklahoma twang, "I'm so scared, I might shit my pants." Her laugh came out more like a squawk. She'd already picked up Jason from childcare and planned to stick around until she heard the committee's decision. "What if I don't pass?"

I gave her a sympathetic look and stuffed a stack of assignments to grade and my lesson plan book into my tote bag. "You and Jace can wait at school if you want," I said, "but I don't know how long it'll be. Why don't you go on home?"

Jason toddled around the desks with his mom trailing behind looking uncharacteristically forlorn. Nikki wasn't an inch over 5 feet tall, but those 5 feet were packed with 10 feet of spunk.

Nothing that came out of her mouth was sugar-coated. She confronted parenting and hardship without complaint. She was used to adversity. Nikki made me laugh; she made everyone laugh with her earthy language and proclamations no one else dared utter. This new challenge, though … I was afraid for her.

At less than two years old, her son was more than half Nikki's height already. He babbled and grabbed at the lab supplies I'd laid out on the tables—too near the edge, apparently—for tomorrow morning's biology lab. Nikki followed behind and replaced what her son pulled down.

"Don't let him get into the bottles of peroxide," I cautioned. "They'll break if he drops them."

Nikki picked her son up, his long legs stretching down the length of her slender body. He grunted and reached for a test tube rack, but Nikki wrenched it free, and then brought him over to my desk. I kept a few toys there that were safe for the babies who came with their moms for after-school detention or during special events. Jason chose one of the Matchbox cars we used in our inertia lab.

I glanced at the clock and pulled my purse out of the desk drawer. "I've got to run. The committee convenes at four. I'll have to hustle to get to the high school and park." I imagined all the students streaming out of the high school parking lot in their BMWs and Audis like jittery cattle through a sprung-open gate, and me steering my old-lady Nissan Maxima sedan against the flow. The MHP parking lot, with its 15 or 20 cars in three compact rows, was easy to navigate. There weren't any BMWs there, but more than one rust bucket—owned by teachers and students alike.

"Try to relax," I said. "You'll be fine." Truthfully, I wasn't at all sure she would be. I wished I had something more encouraging to say, but I didn't want to give her false hope. I assured her I would call with the committee's decision and then locked my classroom door after us.

Nikki sighed and walked down the hall with me to Mrs Sledge's desk, the centrally located hub of the campus, anchoring the education and social service wings on either side. She squared her shoulders and gripped Jace so tightly that he squirmed and whined to be put down. When we got to the end of the hall, Nikki sat Jace up on the counter and greeted Mrs Sledge, turning to give me a slight wave as I left. Mrs Sledge was the school secretary, the principal's administrative assistant, the registrar, the students' bottled water supplier (pusher), the cafeteria chair straightener, the morning announcement emcee, and every student's friend and confidante.

My anxiety level was nearly as high as Nikki's because I wasn't sure how much I could do. As one of only five teachers on a small campus in a large suburban district with hundreds of teachers at the main high school campus—the largest in Oklahoma—my voice counted for little. We were less than two miles from the main campus—out of sight and out of mind—on purpose. Because of our size, students knew every teacher, and teachers knew all the students, even those who weren't in their classes. It meant students were always up in each other's business, for good or ill. At our Wednesday morning staff meetings, the principal gathered teachers, counselors, childcare workers, and nurses to get all up in their students' business, too, but with the goal of eliminating obstacles to their academic success or the well-being of their children.

Nikki came to us as a pregnant junior, almost two years before. She was 19, with a few class failures on her record. She was on an Individual Education Plan due to mild learning disabilities, which meant she qualified for special programs. The small staff and campus meant we couldn't offer all the accommodations available at the high school. Ashley, our social studies teacher who was also certified in special ed, made students aware of all the opportunities provided for them. I taught every student their science courses unless they'd completed them before coming to us. Nikki took two of my classes, one of them for credit recovery.

Although I'd taught Junior English for a few years, Nikki had completed it the year before I started teaching it. She'd passed every class at MHP, but she hadn't been able to pass all her End of Instruction exams. It wasn't that we hadn't tried. A lot of time and energy went into lessons about strategies for passing the exams and reviewing the many practice questions provided by the district and state. I'd used those strategies myself throughout my educational career; I used them for the certification exams I'd taken in 2007 to become eligible to teach. I knew they worked. Teaching test-taking strategies detracted from learning new content, but it was worth it if it helped students pass the exams.

In 2012, educators all over the United States were still in the throes of compliance with the No Child Left Behind (NCLB) initiative. One of the elements of the No Child Left Behind Act was punitive: Schools (and teachers) were punished or rewarded based on student test scores. Oklahoma had responded to the national mandate by creating End of Instruction (EOI) exams for high school students in seven subjects. Students were required to pass four of the seven exams before they could earn a diploma, regardless of their GPA or whether they'd received the required course credits. Two of the four must be Sophomore English (English II) and Algebra I.

At the same time, we were in transition. Common Core Standards were adopted by Oklahoma and most other states in 2010, initiating a headlong hurtle away from NCLB and toward a thoroughly different educational philosophy. Our district hosted workshops, invited speakers, and provided instructional manuals to teaching staff as a means to educate them on the new standards, which were intended to be common throughout the nation. We had no way of knowing our efforts, and the millions of taxpayer dollars spent would be completely wasted when Oklahoma repealed Common Core in 2014 after most of us finally comprehended the framework concept. In 2012, we were still in limbo with one foot stuck in the mucky territory of the EOI high-stakes testing mandated by NCLB and the other in the promising new land of Common Core Eden.

Nikki had taken the English II EOI three times; she'd failed all three times. She'd passed three other exams as required, but her only hope for a diploma was to complete the alternate project only available to special ed students. Because I hadn't taught her English but knew her situation, I was invited to serve on her OMAAP (Oklahoma Modified Alternate Assessment Program) Project Committee. Her project score would substitute for the English II test score.

It was the spring semester of her senior year, and this was her last shot. If she didn't receive at least a "satisfactory" result from the committee members, she'd end her high school career without a diploma. Consequently, her employment options would be limited. The stakes were too high to ignore. I would join two other English teachers at the high school for the certifying committee, teachers who'd never met Nikki, and whom I'd also never met.

What held Nikki back on every one of her failed EOI attempts was the essay portion of the exam. After reading her muddled science lab reports and assignments over the past two years, I was deeply concerned about the outcome of this project, which centered on an essay response to a prompt chosen by the district OMAAP coordinator. According to her current English teacher, Nikki could read passages and perform well enough on questions about main ideas, supporting evidence, and story tension. As an astute observer of human nature, she was spot-on in detecting character motivations. Nikki's language construction skills, on the other hand, were atrocious. Words mixed themselves up in her head and spit out sometimes confusing, unstructured sentences. She had only a moderate sense of where to begin and end a paragraph, much less when to insert transitions. Punctuation was injected randomly, or not at all, to add to the confusion. I was afraid of what might confront me at this meeting. She'd submitted the essay a few weeks earlier, and I'd never seen it. I feared Nikki's project performance might elicit scorn from the other teachers, not only for her but also for our program.

It was no secret that our student body was considered deficient by other district employees. In the past, when I'd met instructors from the huge main campus, I detected almost palpable disdain for our students and, by extension, their teachers. Feelings of inferiority dogged me throughout the entire eight years I taught there. It rendered me essentially mute in every departmental meeting I'd attended, beginning with the one in which one science teacher had remarked that MHP student scores had likely brought down the district's biology EOI average. Quite a feat for our dozen or so students that year among the several hundred at the high school. Nikki wasn't a star student at any level or in any subject, but it seemed there was more at stake here than one student's essay. It felt as though our program was being assessed.

When I arrived in the meeting room, I introduced myself to the two other teachers on the committee, whose names I forgot almost immediately. My mind was blurred by the dark cloud of apprehension. I'll call them Sally and Jane.

Sally, who chaired our small committee, passed out copies of the essay Nikki had submitted for her project, responding to a prompt about her future aspirations. "Nikki completed all the project requirements," she began. "Our job now is to determine which level of proficiency the essay demonstrates. Just so we're on the same page, our choices are advanced, satisfactory, limited knowledge, or unsatisfactory." She passed out a printed description of each level for the English II exam. A satisfactory or advanced score was required for Nikki to pass.

We were quiet for a few minutes, as we each read Nikki's lengthy handwritten response to the prompt, which asked about her accomplishments and goals. Sally and Jane made notes on their copies, but I left mine unmarked. It was a bit painful to read, but somehow it felt like a betrayal to mark corrections. I resented the slashes and comments they scribbled across Nikki's text. My students frustrated the hell out of me at times, but they were *my* students. I'm sure the high school teachers had students like Nikki, but they also had students who wrote insightful articles for the school newspaper, aced their SATs, and qualified for National Merit Scholarships.

Nikki had passed my Environmental Science class as a junior, the year she came to us, fulfilling her optional science credit. She was currently repeating Physical Science as a senior. We were well into the second semester, and she was doing okay. The modifications she qualified for as a special ed student helped a good deal. She might not be college material, but she was on track to pass the class.

Some of the challenges Nikki had faced as a pregnant 16-year-old poured out in her essay. She'd been dumped by Jace's father as soon as she got pregnant and now lived with her son and mother, working when she could get hours and babysitters. She was also one of the few students who had a car. It was a clunker, but it ran, and she provided rides to her classmates when she could, which she enjoyed being able to do. Her use of punctuation made her future hopes difficult to read, but her vocabulary was thoughtful and her argument was well supported. She'd created a safe home for herself and Jace, and he was thriving. She acknowledged she wasn't gifted with intelligence or the benefits of the middle class, but asserted her advantage was her determination. Many of her words were spelled incorrectly, but they demonstrated a mature, confident young woman who was willing to face adversity for the well-being of her son.

Sally and Jane made a few comments about the structure of Nikki's sentences and her writing skills, but they then looked to me for my impression. It was clear they were deferring to me as someone who knew Nikki. "What do you think?" Sally asked.

"I've taught Nikki in a couple of science classes," I began, "and I can tell you she's a determined young mom. She'd do anything for her son. You can see how passionate she is about Jason's well-being." I read a passage aloud from her essay about how important she felt it was to teach her son to value a good education, and I found myself tearing up. I'd often been irritated with students, including Nikki, who seemed not to care about the subjects I needed to teach them. They could be inconsistent in their attendance for a variety of both good and bad reasons. It often bothered me that their grades didn't reflect their true abilities because they were laissez-faire in completing assignments. Yet here was Nikki demonstrating through her words how important education was for her son.

"I know what you might be thinking. There are a lot of mistakes in this essay. Not everything Nikki has learned by being a teen mom can be measured by an End of Instruction exam or how well she punctuates prose," I finally said. "But it's obvious from her essay that she's more mature than the average high school senior, and she's resolved to make a good life for her family. I think she deserves a chance to work toward it. Without a diploma, she won't have many options. Her son won't either."

I hadn't said this much aloud to teachers outside my campus in the five years I'd taught in the district. At every juncture, I sensed that the opinions of others were valued over mine. Their classrooms fit the district norms; mine didn't. They were degreed career teachers; I was first a medical technologist and then a writer before becoming a teacher. They taught students for a semester or a year, wished them well, and sent them on their way. My colleagues and I were serious about our program goals to see that the girls in our classrooms succeeded not only as students but also as parents.

I think I was the only one surprised by my words, which had come without deliberate forethought. But once they were released, I was convinced of their truth. Nikki deserved her chance.

Sally nodded slightly and looked back at the essay. She finally said, "It'll take two 'satisfactory' scores to confirm a passing grade." She smiled at me. "Thanks for adding your personal perspective. I vote that Nikki's project be deemed satisfactory."

I hadn't realized I was holding my breath until then. "Thank you," I said, finally exhaling and looking down to hide damp eyes.

"I also think Nikki should receive a satisfactory score," said Jane.

"Well, it's no secret how I vote," I said, grinning in relief. Both teachers laughed. "You have no idea how important this is to Nikki. She's been a nervous wreck all day and is waiting back at school. Can I call to give her the good news?"

"Of course," said Sally, and she promised to file the appropriate documents.

As soon as I got in the car, I called Mrs Sledge. When I delivered the verdict, she repeated it aloud for the benefit of anyone who might be near her desk. "You passed," she said, and I heard Nikki's scream in the background.

"I'll be back at school in a few minutes," I said. "Tell Nikki congratulations for me."

When I pulled into the MHP parking lot a few minutes later, Nikki's car was gone. Only Mrs Sledge was left, and she met me at the door with a hug. "Nikki was in tears when she left," she said, "but she was in a hurry to tell her mom. She said to thank you."

I'd begun teaching at MHP in 2007 not knowing what the hell I was doing, and I limped through the first few years thinking nothing I did made a real difference in the lives of these girls. I felt like choking or hugging them, depending on the day, but I never did either. I was adamant about keeping a professional distance. Only one or two had come back to visit since they graduated, and I had very little idea who'd fulfilled my hopes for them and who had not. Before serving on Nikki's project committee, I didn't know how it felt to so directly affect a student's success. I've since learned such occurrences are rare, and this one I treasure. It was all I'd hoped for when I first stepped into the classroom as a middle-aged, inexperienced, science-nerd, wannabe teacher.

16
Technophilia: Don't place too much faith in structures alone; technology is a means to an end

Right in the middle of the fall semester's online final exams in 2012, Poof. The power went out. It was gray and cold outside—freezing, actually—on this day during the last week of class before winter break. Even before this, students and staff were distracted. Everyone looked beyond finals to holiday celebrations, including me.

"What do we do now?" the girls asked as soon as the lights went off and the Wi-Fi went dead.

"Let's give it a few minutes," I said, trying to be calm despite a surge of anxiety. I glanced at the clock. Every minute the power was out meant one less minute of productive time, and there was never enough of that. I'd been faithful in adhering to Mrs Coleman's advice to keep students engaged bell to bell. I didn't have time for this kind of interruption. Even if power came back quickly, we'd have to restart the modem, and everyone would need to log in again. I couldn't be sure their answers had been saved.

"Let's just cancel finals," one girl helpfully suggested.

"Let me see if it's everyone or just our class," I said, before looking out into the hallway. The hall was dark and quiet. Ashley from next door came out as well.

"What do you think?"

She shrugged. "Wait and see, I guess."

***

techne – art, craft, skill, or later: technical/technology (Greek)

phile – one that loves, likes, or is attracted to (Greek)

These two terms appeared on my weekly word roots assignments for all my classes sometime in 2010 or 2011, but I never put them together. I'm not sure technophile was even a colloquial word in 2010. Until the twenty-first century, maybe there was no need for it.

In a 2001 article titled "Digital Natives, Digital Immigrants" in *On the Horizon: The International Journal of Learning Futures*, Marc Prensky coined terms for those who were born before the advent of personal computers and the internet (digital immigrants) and those born after (digital natives). The implication is that those born before 2000 are much less likely to adopt or be comfortable with digital technology than those born later. Others have used the terms "technophobe" and "technophile" to describe the same attitudes. Prensky also postulated that this difference in characteristics between the generations created the greatest challenge to education. There's some merit to his thesis, but it's far from clear-cut. The students I taught between 2007 and 2015 were born between the mid-1990s and early 2000s, which means they straddled that hypothetical line between digital generations. All our students had cell phones, but most just used them for texting, scrolling social media, and the occasional calculation in math class. Maybe they'd use GPS to navigate.

I'd taken a graduate school course in the mid-1990s in online database searching, and while the internet was available then, it wasn't widely accessible in homes. I was captivated by the concept of finding so much information without a trip to the library and flipping through a card catalog or magazine stacks to complete a research paper, which essentially every course required. The online database course project introduced me to topical academic databases, which ran the gamut from health to business, and popular culture to psychology research. I predicted at the time that all knowledge available to humanity could someday be reached digitally. That prediction has essentially come to pass. There are still databases that are only accessible through academic institutions, but for the most part, if I have a library card, I can access almost any information I need. From home. In my pajamas. I was hooked. Did that make me an ahead-of-my-time digital native, or a just-in-time digital immigrant? By the time I started teaching in 2007, I'd been writing and editing on the computer for a decade. It was exhilarating to have so much information at my fingertips! Of course, it also introduced me to rabbit holes of research into which I could descend for hours. No human innovation comes without a downside, I suppose.

One of the advantages of teaching on a small campus in a large district was that we could test-drive initiatives. We were the first campus in the district to get Smart Boards. I fell in love with mine. The reception was mixed in other MHP classrooms—not all staff were as enthralled by technology as I was. Overall, though, it was judged a valuable teaching tool, and the district eventually installed them widely. Over the next few years, our campus would go on to pioneer the concept of computer-equipped classrooms and then a program—now the custom in most public school classrooms—in which students were issued a laptop for their use throughout the year.

In 2010, we won a grant from a local internet provider for a cart of 20 laptops, which meant teachers had to reserve them for class use. I admit to being a laptop hog; I loved having access to so many online resources. Soon, I became frustrated by the sign-out process and petitioned for the privilege of keeping the cart in my classroom. It was purely selfish, I know. While other teachers used them, I planned most classes around online lessons. I liked having control over ensuring they were charged consistently and went through the cart each afternoon before leaving to plug in those that had been hastily thrown into a slot.

One of the primary reasons we received the grant was because each of our moms spent two to three weeks on maternity leave during their stay with us. The girls checked out a device to complete their assignments while on leave, so they weren't too far behind the rest of the class when they returned. This was a mixed blessing for students, of course. We'd always sent assignments to them during leave, but we'd previously been limited to textbook exercises and printed quizzes. Once students had access to a laptop, they had no excuse not to keep up with the (in my opinion) more engaging online work. Not all students had internet access at home, so we partnered with a local internet service provider who offered access for an affordable monthly fee to low-income families. By then, there were also a great many public access points in the community.

In 2012, after our limited laptop cart trial, the district invested in new laptops for our students in a one-to-one pilot program. Shortly before this, I took on the role of Lead Instructional Technology Trainer for our campus, conducting training sessions for staff on new district technology. By this point, I was steeped in all things digital. I was doing most of my lesson plan research online and making use of digital resources as much as possible. Because all students now had access to, and responsibility for, laptops, I created an online Digital Citizenship course. They were required to pass the lesson, "Need a Digital Passport?" before they were assigned a laptop, something many students resisted.

Eventually, they all passed the test and were assigned a laptop, bag, and charger. Perhaps students didn't want the responsibility of caring for one at home. We'd made it clear their parents would pay for lost or stolen laptops. Or perhaps the girls just didn't want constant access to a device that was required to complete their assignments.

Around this same time, I began using a shiny new thing, an online Learning Management System called Haiku (which has since become PowerSchool) to post lessons, link to resources, and create assignment submission links. They were integral parts of my lesson planning; I no longer used a written lesson plan book. Instead, I printed the lesson descriptions to turn in to our principal each week, as required, and created access links for students and parents in Haiku. I spent my weekends filling each class's content in the Learning Management System, which absorbed most of my weekend but saved a lot of time at school and made it unnecessary to transfer files from home to school via Dropbox, which was becoming cluttered with files. I was in heaven.

My students? Not so much. I think students resented that they could no longer claim they didn't have the tools they needed to complete work outside the classroom. They were frustrated by the necessity of learning technology on top of learning science. I was surprised and disappointed they weren't as excited by the laptops and Haiku as I was. While I took Prensky's labels of digital natives and digital immigrants with a grain of salt, I assumed that because technology would clearly drive their futures, my students would embrace it. Some staff resisted as well, and I wasn't surprised by their reactions. However, I was dismayed that so many students failed to accept the technology we provided for them. They were careless with laptops and groaned when I demonstrated the digital resources they were required to use to complete assignments. I'd expected to align myself to their much more facile digital usage only to discover that I was more technophilic than they were. I genuinely loved the benefits of technology in education, but most students didn't share my enthusiasm.

The next few years, which became my last as a high school science teacher, were made more stimulating by shiny new digital tools. I dug into those innovations with eagerness, because

I enjoyed the challenge of learning how to use them. I realize now that students had more pressing issues; their goals differed from mine. My goal was to present information in ways that engaged us all. Finding activities created by professors, professional scientists, or other high school teachers from around the world broadened my approach to the sciences I taught. I accumulated a vast number of resources over the next several years before leaving MHP in 2015.

When a middle school teacher was hired to replace me, she asked for any help I could provide. I gifted her a thumb drive with copies of hundreds of lessons and resource files. I packed up three boxes of science books full of activities, labs, and lessons for my district's Science Coordinator. I preserved my Dropbox folders of lesson material until I was able to copy my MHP files onto an external hard drive. In eight years, I accrued thousands.

Now, when I scan the dozens of folders and thousands of files and consider the investment of my energy they represent, I'm convinced I don't fit either of Prensky's definitions well. I'm neither a native nor an immigrant. Maybe I'm a naturalized immigrant, or what I'd now call a technophile. However, I'd like to propose an additional term that Prensky never considered. I think I may be a digital hoarder, a condition for which no 12-step program promises a cure. As I scan the voluminous contents of those eight years of files duplicated via three current archival strategies: thumb drive, my newest laptop's C drive, plus an oversized external hard drive, my eyes tear up … just a little. Each file represents the hope for a better way to deliver a lesson, a more graphic illustration of a concept, or a link to cutting-edge research data. I like to think my backups protect me against catastrophic loss. I'm an unrepentant digital hoarder, which sounds dangerously close to a psychological diagnosis. I don't care. Only a devastating failure of yet unknown origin will pry those files out of my cold, dead digital drives.

***

In the dark and quickly chilling classroom full of online test-takers in December 2012, my students and I pulled on our coats and gloves. A few brought their babies back to the classroom, not because it was any warmer there but to share body heat.

"Let us go home," they begged.

"Surely the power will be back on soon," I said, though I wasn't at all sure it would be. Not all our students had cars, and I suspected if we canceled classes, the power, heat, and internet would be restored before their rides appeared.

I honestly don't remember how students completed their exams during the winter storm internet outage of 2012. Other classes were administering paper and pencil tests and had enough sunlight to take them; mine depended on the internet connection. When I asked Ashley, she remembered power being restored at some point that morning but not how long it was out. Perhaps I chose to let grades stand where they were before finals. That would have been reasonable, and I have a vague sense that this is what happened. All I know is that the incident taught me not to depend so completely on technology.

By the time classes resumed in January, I was a bit more tentative in my insistence on using the online submission links for assignments. I learned to be a bit more flexible in accepting neatly handwritten work when there were problems with online access. I learned to appreciate the wise teaching maxim I'd heard years before about planning more material than you expect to need— especially alternate material. The idea has applications beyond digital technology, too.

I still consider myself a technophile and often become enamored with a new digital tool. I'm just more selective about which ones I fall in love with. Instead of being driven by someone else's instructional calendar, I can now choose technology platforms to satisfy personal interests; I no longer grasp every shiny new thing.

17
Controlled variables: Recognize biases and work to overcome them

Almost as soon as I pulled the book out of her hands, I suspected I'd made a mistake.

Sophie glared at me. "You shouldn't have done that," she growled. "You can't just take my stuff away from me."

I'd seen this picture book in childcare; I knew it didn't belong to Sophie. I also knew she hadn't heard any of the instructions I'd just given for the class assignment. "You need to focus on the lesson," I snapped. "I asked you to put the book down. Pay attention."

Instead of backing down, Sophie didn't let it go. She was angry. So was I.

"Give it back." She put her hand out, but I reached beyond her to drop the book on the edge of my desk.

"After class," I said.

She pulled it from my desk, challenging me with her eyes and a slight smile. Our eyes locked for a few moments before I made my second mistake of the day.

"Come with me," I said, motioning for her to follow me. "Let's go see Mrs Coleman."

By this time, I'd been teaching science for several years and had quite a few tools in my teacher toolbox. I'd taught a lot of students who were angry, confounded, or overwhelmed by a pregnancy they hadn't counted on, but I'd never encountered a student like Sophie—a student who so openly defied authority and pushed every limit. She upset my hard-fought composure with her assertive self-confidence. From the time she enrolled, the stepsister of a former student and the half-sister of another who enrolled with her, her attitude touched off something in me as no other student had. I just didn't respond well to Sophie. She unnerved me.

One of the elements of the Scientific Method that I drilled into students from the first week of class was the concept of a controlled variable. It differs from dependent and independent variables in that it remains the same throughout an experiment. Controlling the temperature or the light at constant levels as you test the effects of varying amounts of water or fertilizer on plant growth are examples of controlled variables. Keeping your cool when confronted with a defiant student while testing student-led instructional strategies in biology class might be another.

I prided myself on not overreacting to student behavior, pretending to take everything in stride. I'd internalized the adage to never let a student see you sweat and thought I'd honed the art of coolness in the face of sarcastic comments. Sophie tested that hypothesis. An accurate hypothesis depends on predicting what effect the variables have on the outcome of an experiment, though. I couldn't have predicted Sophie's outright refusal to take direction or her negative responses to most efforts to engage her. I also failed to take into account her complicated family history.

Our principal restored peace between Sophie and me, establishing that the book we sparred over should be returned to childcare after class and that Sophie should spend science class time on science. Mrs Coleman's picture should appear alongside the definition of "optimist" in the dictionary. She could convince anyone they were capable of reaching the high expectations she set for them—and her expectations were high for teachers as well as students. Sophie and I left her office both feeling we'd won the skirmish. Nevertheless, I had the distinct impression that Mrs Coleman concurred with Sophie that I shouldn't have just pulled the book from her hands. There was a nagging sense that bringing my dispute with Sophie to Mrs Coleman was akin to the squabbles my children often brought to me for resolution when they were young. It left me feeling unsettled and somewhat defeated.

***

Later that spring, Mrs Coleman asked me to coordinate a campus Career Fair. Many of our students came from families where parents and grandparents had never attended college. Their ideas of careers were often limited to the professions they'd personally witnessed, narrowed by their youth and circumstances. I hoped highlighting more varied careers would expose the girls to broader options than they'd considered. Most of them leaned toward futures in medical or dental assisting, childcare, or cosmetology. The local vocational-technical school, Tulsa Technology Center, provided training that allowed juniors and seniors to spend half of each day on their campus and half on ours. Many graduated from high school with a certificate in hand for entry-level jobs. We encouraged every avenue of education beyond high school, including the technical programs, but we also hoped more rewarding or higher-paying jobs would inspire some.

I arranged with area college and technical school representatives to set up display tables in our cafeteria on the afternoon of the Career Fair and enlisted several local employers to provide presentations about their businesses in morning breakout sessions. The Tulsa Chamber of Commerce gave a presentation on personal finance. We suspended classes for the entire day but required attendance and asked the girls to dress as they might for a job interview. The most professionally dressed student would receive a gift card to a local clothing store for her efforts.

In the morning session on finances, the speaker asked each student to select a strip of paper from a basket that would reveal her hypothetical job and annual salary. As a table, the girls then discussed the expenses printed on the handouts they'd drawn, such as rent, utilities, groceries, as well as extras like restaurant meals, electronics, and fashion items. Each student created a budget, choosing which of the expenses she could afford on her salary. The point was to learn how to better manage money, and to introduce students to realistic expectations for their intended careers. It became clear that the cosmetology and medical-assisting jobs some aspired to wouldn't allow the girls many common luxuries unless they advanced to owning their own salons or later getting a nursing degree.

"What did you learn?" the speaker asked after students had discussed and created their budgets. After a few students made comments, Sophie spoke up.

"I drew 'retail business owner'," Sophie said, waving her strip of paper in front of her, and stated the annual income on her sheet. She cocked her head to the side and let slip one of her unique smiles. "It's more than the maximum income that qualifies me and Colby for state benefits. If I made this much money, I'd lose my benefits and would have to pay for childcare out of pocket. If the business I owned was beauty products or shoes, I'd have a lot of fun," she said, smirking at all the girls at her table, "but I'd actually lose money."

The speaker sputtered a halfhearted response about educational goals and career growth before calling on another student. I was irritated. Sophie's awareness of how much money she could earn while still qualifying for state benefits was a testament to her shrewdness, but the attitude it implied appalled me. Her convictions about self-reliance and work ethics were vastly different from mine. Perhaps that was the source of the tension between us.

Sophie was smart enough to hold a decent job and provide a secure life for herself and her son. There was nothing deficient in her intellect. However, if she could obtain benefits that provided for basic survival without working at a better-paying job she was perfectly capable of, she was more inclined to take that route. It was disturbing that Sophie's future employment plan hinged on maximizing her public assistance benefits. Social support programs fail many of the people they're designed to help, and Sophie seemed determined to suck resources out of a beleaguered system without considering how it would further jeopardize those whose existence depended on it—including student peers who weren't gifted the physical and intellectual benefits I believed Sophie had.

***

The summer between Sophie's junior and senior years, Mrs Coleman recommended her for a scholarship to a local program that trained young people to become certified nursing assistants. I sometimes felt Genell saw more potential in students than I did, but I was still surprised at her assumptions about Sophie. While I knew Sophie was intelligent enough to succeed in a nursing assistant program, I wasn't sure she'd view it as the opportunity Genell did. To Sophie's credit, she accepted the scholarship and completed the program; she received her certificate, making her highly employable albeit at a relatively low-wage job. I never doubted she was capable of obtaining the certification and suspected the salary wouldn't be high enough to bump Sophie's eligibility for benefits. Perhaps it really was her best option.

MHP staff, myself included, always wanted the best for our students. Our personal expectations could be draining, though. Teacher burnout is often attributed to the physical demands of the job, which are significant, but the emotional exhaustion of sustaining high hopes for students is underrated. I was genuinely proud of Sophie's new certificate and told her so, but after a half-dozen years in the classroom, I was learning to temper my hopes.

Sophie graduated in 2014, a year before I left my position at MHP, and a few years before the program closed its doors permanently. I'd heard that the boy she was living with when she graduated was arrested for drug possession that same year, and I wondered if Sophie would have similar run-ins with law enforcement. But I didn't give her much thought once she was gone. I went on to a couple of other part-time assignments for Tulsa Technology Center and continued teaching as an adjunct professor at Tulsa Community College. I had little time to think about this one student—this one young woman who stretched the limits of my patience and composure more than any other.

***

About a year later, I spotted Sophie across a crowded and noisy restaurant one Saturday evening in Tulsa. My husband and I were having dinner with out-of-town friends who were visiting for the weekend. As soon as I recognized Sophie, I turned to focus on our table's conversation. I didn't glance back at Sophie, but I was very much aware of her presence. It dredged up feelings I'd thought had been forgotten. I hoped she wouldn't notice me because I wasn't sure what I'd say to her. Her presence still had disturbing power.

After our bill was paid and we stood up to leave, I heard Sophie call out, "Mrs Airhart!" When I turned, she was there behind me, flashing a genuine smile with arms open for a hug, and I put my arms around her.

"How are you?" I asked, and she said she was fine. She babbled a little about what she'd been doing in nonspecific terms. I was tempted to ask, "Are you working? Are you being a good mom to Colby?" Instead, I told her it was good to see her, and she responded likewise.

As I left, it occurred to me that Sophie didn't have to leave her table or her friends long enough to greet me, but she did. I hadn't been willing to do so, but she had. In the years since this chance encounter, I've turned it over in my mind and wondered what it said about her, and what it said about me. In some ways, Sophie's hug troubled me more than the classroom tensions we'd stood on opposite sides of for a couple of years. It bothered me more than the snarky and self-satisfied attitude I attributed to her. Maybe I'd misjudged her. Maybe Sophie was simply a young woman who hadn't ever had a supportive influence in her life, and she'd responded with the same bitterness and disrespect that she'd been raised with. From the distance of years, I now suspect this is true. If I were honest, my reaction to her hadn't demonstrated the compassion I expected to feel when I started teaching. I was disappointed in myself. Perhaps what irritated me most about Sophie all along was that she made me face my own shortcomings; she made me recognize that I'd let my personal biases about appropriate behavior determine which students deserved my compassion and which didn't. I assumed one of the controlled variables in my teaching career was my sincere desire to help all students reach their potential. I wanted this to be true. Sophie made me reevaluate not only my hypothesis about teaching but also the conclusions I'd drawn about us both.

18

Momentum: Newton's Second Law: Don't underestimate students' abilities

In late 2012, I found myself in a position to purchase some badly needed laboratory supplies. MHP received a small allowance from the district each year to spend on classroom resources. Split between five teachers, my portion wasn't all that much, but it was still crucial. That particular year, for unknown reasons, none of the other MHP teachers submitted requests. Mrs Coleman reminded us several times of the deadline, but I was the only one who turned one in. There was no way she was going to leave that money on the table, so she informed me on Friday afternoon that I could use the entire school's allotment if I got my revised request to her by Monday morning. I did, and it was all approved.

> *Note to district administrators*: Approving a lump sum disbursement to a science teacher for lab supplies is akin to dropping a chocoholic with a spoon into a vat of molten Godiva dark chocolate raspberry truffle.

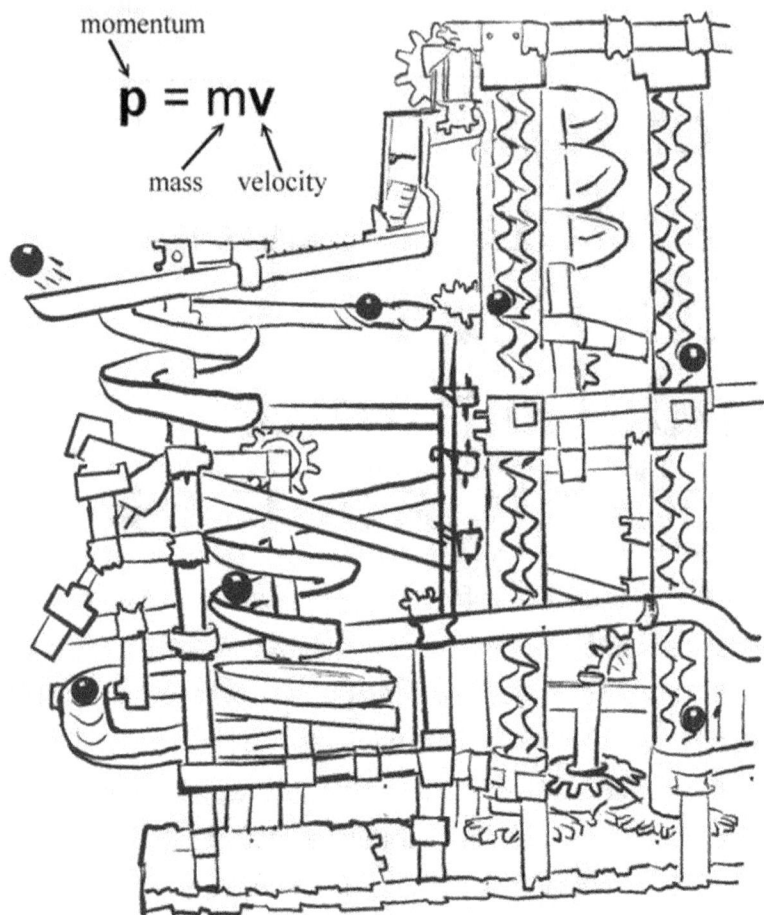

momentum

$$\mathbf{p} = m\mathbf{v}$$

mass   velocity

I don't remember all my choices, but I do recall ordering some basic art supplies that were used throughout the year for various poster projects. I bought a few of the lab supplies I'd been in the habit of buying myself and knew I'd need. One purchase proved quite popular: the large and extra-large exercise balls in blue and pink—two per classroom—for students to sit on during class. I think buying them for every classroom made me feel less guilty for hogging all the funds. Sitting on inflated exercise balls alleviated some back pain for students in the later weeks of pregnancy. The blue plastic chairs with the chrome legs the district issued to our campus were uncomfortable, even for non-pregnant students. I didn't much mind the bobbing and weaving that ensued, but settling the arguments over who got which ball became tedious. There was trading between classrooms, and when a student laid claim to one as "hers," bickering was inevitable. At the end of school days, teachers would retrieve one or more from another classroom where a student had carried it, insisting she could sit on nothing else.

My most memorable purchase—one that seemed a tad frivolous at the time—was a kit called "Marble Mania Extreme." At just over $100, representing a sizable chunk of the funds I was allowed, it seemed somewhat of a gamble, but I was a little desperate to keep my protégés from nodding off during a challenging physics unit on Newton's three laws relating to force, motion, and momentum.

My background is in the life sciences, with a heavy dose of chemistry thrown in. I'm not a big fan of physics. Don't get me wrong; physics is important. I've witnessed the untoward consequences of defying Newton's Laws a time or two. After all, according to a poster which hung on the wall beside my desk: "Gravity. It isn't just a good idea. It's the law." I'd obtained a signed copy of the poster from the artist himself, Gerry Mooney, and it always made me smile. However, most ninth graders don't readily buy into the necessity of physics instruction, just as I hadn't when I was their age.

One of the reasons I loved science was because of the hands-on learning that lab activities provided. I'm a strong proponent of kinesthetic learning. No matter what a student's primary learning style is, getting their hands on things is powerful. In previous years, I'd created physically active competitions to illustrate principles of motion.

Despite my indifference to physics pedagogy, I found it created a lot of opportunities for active lessons and demonstrations of observable phenomena. The mathematics of using formulas to calculate aspects of motion was always a challenge, but students really dug into the hands-on portion of labs. It pulled interest better than lectures. This would never have been possible at the large high school most of my students transferred from, where classrooms were packed with 30-plus students as classes at North Intermediate had been.

Marble Mania Extreme seemed a reasonable solution to the problem of showing abstract principles like inertia, momentum, and energy. To my dismay, the contraption arrived disassembled in a 24"x36"x5" box that proclaimed: "501+ pieces!" I'm going to be spending a lot of evening and weekend time on assembly, I thought. Oh, boy.

I couldn't have been more wrong.

Almost as soon as we spread the plastic baggies of multicolored parts on the tabletop and dove into the 57-step assembly instructions, the students jumped in to take on roles. Some specialized in locating parts (thank goodness each part was numbered!), some in interpreting directions, and some in assembly. Most days, they forgot I was in the room. During their Academic Options hours on Mondays and Fridays, students asked to come work on assembly. Hanna took a particular interest in the Marble Mania Extreme. I sometimes tried to help her when she got stuck, but I just slowed her down. This ninth-grade student had also been fascinated with the Engino building kits I purchased earlier that year. I suggested she might someday be an engineer, to which she rolled her eyes.

"Come on, Ms Airhart," Hanna said.

"No, really," I said. "You're good at building things." She just shook her head. It turned out she was right—she wouldn't become an engineer. However, little more than a decade later, Hanna graduated with a Master of Social Work degree from the University of Oklahoma and began her career as a licensed social worker. I'm proud of Hanna's resolve, not only in the classroom but also in providing a loving home and being a superb role model for her daughter.

It didn't take long for Hanna and a handful of other committed students to complete the Marble Mania Extreme assembly. They placed it on a table in the hall outside our classroom for all the school to enjoy. My students were quick to point out which parts they'd completed or show other students how to place the marbles correctly on the conveyor so they'd trip the nifty sound switch on the left side.

I'm still not sure if these students can articulate Newton's Second Law or the formula for calculating the acceleration of a 12 mm marble down a 30-degree inclined plane (I can't do it without cheating), but they learned the most important principle. The higher the marble is placed, the greater its velocity and momentum. They learned how to read two-dimensional instructions to construct a three-dimensional model. They learned that if they stick with a project, even when the purple ramp doesn't line up to the green wheel until they (eventually) turn it around and right side up, they'll have something impressive to show for their efforts. All valuable life lessons.

According to Newton's Second Law of Motion, the force of momentum is the result of an object's mass times its velocity. When two bodies are moving in the same direction (say a teen mom and her baby moving toward self-sufficiency), the total momentum is a product of the two. They will continue in the same direction unless acted upon by an external force moving in a different direction. However, if said teen mom discovers her baby's father has fathered a child with another girl, some momentum is lost to the external force. In this case, the teen mom and her child will lose the energy of momentum that is transferred to the new child. Friction is an unavoidable opposing force dragging against motion as well. The dilemma is whether the teen mom and either or both children are left with enough momentum to continue forward movement.

Our program's goal was to conserve our students' forward momentum—or even better—to propel them forward with greater force. We hoped to boost both moms and babies with increased momentum toward achieving their goals. The Marble

Mania Extreme taught me not to underestimate students' abilities. I also learned that taking a chance now and then, even with an off-the-wall, somewhat expensive purchase of a 501-piece assembly project that challenged students in novel ways, might just be the critical external force that would move them on to more fulfilling futures.

19
Bernoulli's Principle: All students can learn if they're given enough encouragement

Imagine this: a gymnasium full of middle schoolers, raucous and jittery, yet focused on the next trick up the sleeve of their guest speaker, physics professor Jerry McCoy. He pulls only two items from the cart he wheeled in with him. Students lean closer to see what he'll do with them.

"I use a high-power shop vac blowing air out at a roll of toilet paper. As it blows over the top of the paper, the force lifts the paper up into the air stream and blows it out. It shoots a whole roll of toilet paper up about 15 or 20 feet and unrolls the whole thing in four or five seconds."

What's more likely to entrance several hundred adolescents than a roll of toilet paper being shot up above their heads and unrolled in a matter of seconds?

McCoy's highly popular display of unrolling a toilet paper roll into the air above the heads of students in just a few seconds is an example of Bernoulli's Principle, named for Swiss mathematician Daniel Bernoulli. He'd observed that an increase in the velocity of a stream of fluid above a surface results in a decrease in pressure beneath it. While air is a gas, it acts as a fluid in most ways. Bernoulli's Principle applies to objects in midair and explains the possibility of flight due to air's fluid characteristics. In the case of McCoy's demonstration, the air from the shop vacuum blew across the top end of the toilet paper roll and decreased pressure underneath the paper, causing it to fly up into the air and off its cardboard tube.

I didn't have the equipment necessary to carry out this same experiment in class, but I played YouTube videos for students when we studied the principle during physical science class. It loses a lot of its magic on YouTube, unfortunately. A simple classroom demonstration of Bernoulli's Principle I did use involved blowing forcefully across the top of a toilet paper square that's held at the edge in one hand. The force of air across the top side of the paper decreases pressure on the underside of the free edge and causes it to rise as though flying. It's the same principle that makes airplanes or other objects fly when the speed of air across the tops of their wings is fast enough. My lungs don't have enough power for more dramatic displays.

***

I wasn't thinking about Bernoulli's Principle in April 2014, when a perfect opportunity came up to show the girls the effect in action. Mrs Coleman asked me to organize our campus Career Fair again that spring. I'd planned these events for a couple of years already, so I had some contacts to help me get started. Nonetheless, I tried to find new professionals, vendors, or speakers each year. I loved learning about new careers and employers as much as my students did, and I enjoyed the prospect of helping them expand their imagined career paths. I'd spent decades stretching my own imagination and wanted my students to experience the same sense of satisfaction it brought me.

At each year's Career Fair, we planned a session or two first thing in the morning for the whole school: a keynote speaker, a fashion show of appropriate career wear, or a demonstration (how to change a tire, for example). Then there were several breakout sessions in classrooms with various business professionals or employers. In the afternoon, we'd push the large round tables, high chairs, and plastic student chairs to one end of the cafeteria and set up smaller tables for college or career representatives during the last hour of the day.

One of my medical career contacts referred me to a Tulsa Life Flight supervisor, who was receptive to the idea of participating in our Career Fair. Life Flight provided helicopter transport from accident sites or between medical facilities for critically ill patients in the area. I suggested we arrange a table in the cafeteria during our afternoon college fair time so that nursing and Emergency Medical Technician staff could talk with students about Life Flight. I'm sure he received a lot of requests for demonstrations or talks to area residents about their services, so I was pleased when he offered to bring the Life Flight helicopter to campus. I assured him we had enough space on campus to land it, hardly able to believe my luck.

"I can't promise it," he said. "It depends on what emergency calls we get that day, but if it's available, I'll send it over with some staff who can talk with your students and give tours of the helicopter."

Girls always enjoyed our Career Fair days. Just like staff, students appreciated occasions when classes were suspended for a day. I planned enough speakers so that students could hear one or two possible professionals speak during each of three 45-minute sessions throughout the day. We also began with a couple of speakers for the entire student body. They discussed topics like job interviews, budget planning, time management, and dressing for success.

Most of our students aspired to nothing more than the jobs they'd seen family members in or careers their limited experience had shown them examples of. Because they were pregnant, they spent a great deal of time with medical personnel. Nursing and allied health fields are disproportionally female too, so they could see themselves in those roles. A lot of our students took classes at the local vocational school in the medical assistant or practical nursing fields, or associated ones like dental hygiene or pharmacy technician, where they could gain a certificate at or near high school graduation. All are excellent fields in high demand, but some of them don't pay much. I wanted our students to aspire to more, and I didn't give up easily on the futures I saw for them.

The goal of the Career Fair was to offer different views of the confusing world of work, where students could get a glimpse of the work life of a business executive, a public schoolteacher, or an artist, for instance. Inviting Life Flight was related to most students' basic understanding of medical fields but provided a unique area of medicine for them to consider. Beyond that, having the helicopter land in our schoolyard, just a dozen yards outside the parking lot, was the most exciting event all year.

We were busy browsing the college display tables set up in the cafeteria that afternoon, and I kept looking at my watch. The Life Flight contact had agreed to come around two that afternoon if they were able, but I'd heard nothing from him during the day. I was beginning to give up hope when I heard the *Whoop, Whoop, Whoop* of a helicopter that appeared to be getting louder. The hair on my arms stood up, and I alerted the girls around me.

"Listen," I said, holding the arm of the student nearest me. We stopped and listened to the sound grow louder. "The helicopter's coming. Let's go see!" I yelled for everyone to come outside.

When we got out of the front door, we could see the helicopter angling in our direction and moving closer. As it drew near, it hovered a bit overhead before situating itself parallel to the parking lot and landing about 20 feet beyond it. By the time the pilot shuttered the motor and the rotor blades stopped, most of the students, staff, and college reps were in the parking lot watching. Even the customers at Long John Silver's seafood restaurant at the edge of our property had come out to see the spectacle. The girls pulled their hair back and yelled at each other over the noise. Childcare staff brought some of the toddlers to their moms, and the kids pointed, squealed, and kicked their legs in excitement. A couple of Life Flight staff hopped down, and I greeted them.

"Thank you so much for coming! Just watching you land was amazing."

One of the EMTs laughed and spoke to the girls who'd clustered around, some with their children in their arms. "We're glad to be here," she said, then spoke directly to one of the children, "Would you like to see inside?"

Before long, there were lines to get into the cockpit and have pictures taken. I probably took a dozen or more on whatever cell phone I was handed. The back doors were also opened to reveal the mini-emergency room in the belly of the helicopter. We were invited inside to get a quick presentation and demonstration of the life-saving equipment aboard. Kids and students crawled all over the stretcher, mesmerized by the staff, who identified the medical paraphernalia and explained how it was used, all in terms we could understand.

The yearbook that year featured a photo page of babies in the cockpit, EMTs posing with students and babies, EMTs holding babies, and the crowd surrounding the helicopter that spilled over into our parking lot, watching the whole scene. Just looking at the picture gives me goosebumps with the remembered thrill of hearing and seeing the landing and takeoff of the most dramatic visitors to our Career Fair. When Mrs Coleman retired a few months later, a new principal was assigned. Our new principal chose to bus our students to Broken Arrow High School's Career Day instead of hosting our own campus event. I'm glad this memory is the first and last one I think of when remembering the Career Fairs at MHP.

***

The day after the Career Fair, I drew rudimentary pictures of a helicopter on the whiteboard in my classroom and explained to my physical science class that flight of any kind exhibits Bernoulli's Principle. I wrote the name next to the helicopter. I'm a terrible artist, but my awkward drawings seemed to generally get my points across.

"When the rotor's blades spin fast enough, they work just like the experiment we did in class of blowing over a sheet of lightweight paper," I said. I held a piece of paper in my hand. "See how the end opposite my hand flies up?" I puffed air across the paper several times so they could see it. Then I drew some arrows across the blades I'd sketched on the board to illustrate air flowing over them. "The faster these blades go, the greater the difference between the air pressure above and below the blades. When it's great enough, the air pressure drops below the blades to lift the helicopter off the ground. The same principle applies to birds, airplanes—anything that flies."

Physical science, which included a semester of basic chemistry and a semester of basic physics, is taught in Oklahoma to ninth graders. I've always thought that was a mistake. Without having completed Algebra, their mathematical skills weren't sufficient yet to make the formula calculations required. In addition, many physics concepts were abstract. I struggled to explain concepts like forces, motion, atomic structure, and electricity. Bernoulli's was a specific and detailed concept that had been difficult to get across.

After students witnessed the Life Flight fly-in, there was a dramatic image to connect with this strange name. Maybe none of those students could tell you the name of the principle involved today, but I'm sure they recognize there are scientific principles of air pressure involved in flight.

I used this analogy once in a conversation with Sylvia, the half-time math teacher who shared my classroom and faced teaching concepts even more unpopular than science. When she voiced her disappointment in some of her students' math performance, I responded, "We get these girls whose understanding of math or science is here," and I held my hand up about waist high, "and we take them up to here," moving my hand up to mid-chest level. "They may not get where they should be," I said, again moving my hand up to about chin level. "But they leave us at a higher level than before they came."

If we blow enough "air" over their wings, in the form of creative lessons and caring encouragement, every student will rise. They may not reach the level of a Life Flight helicopter, but they will rise.

20
Tectonic shifts: Meeting students where they are promotes trust

"Blue's my color, don't you think?" My student held a *Seventeen* magazine spread next to her face, showcasing a satiny, azure blue strapless number. She waggled eyebrows above eyes a shade or two lighter than the dress.

"Ooh, that one's sexy!" Her classmate said, cocking out a shoulder.

"Girls." I was losing what little patience I had and signaled for the girls to put the magazine away.

"Tectonic Plate Theory describes how different boundaries form between the plates of land mass 'floating' on the molten layer of earth below," I'd explained, just before class was disrupted. "The boundaries are named for the direction of movement. Divergent, convergent, and transform boundaries form where plates meet. Landforms like mountains, or disruptions like volcanoes or earthquakes form at the boundaries."

We were halfway through a tectonics lesson when I made the mistake of answering the ringing phone on my desk. The campus

nurse called for one of my students to bring her baby for a well-baby check, so she gathered her purse and left for childcare in the next wing. By then the class had devolved into clamoring over glossy photos of off-the-shoulder gowns, strappy, sparkly, sandals, and updos. I don't know where the *Seventeen* had been stashed, but when I picked up the phone, the girls whipped it out faster than a starving lizard snags a fly.

There was only one explanation for this giddy searching for the perfect dress and the perfect hairstyle, reserving a table for dinner, booking the stretch limo, and negotiating who would ride in it. There was only one explanation for these furtive (and not so furtive) preparations interrupting every conversation on campus over the first half of April each year.

Prom.

***

When a girl enrolled at MHP, she was often in shock. Becoming a mother at 13, or even at 17, was rarely an adolescent's plan. Adjusting to what followed could derail her vision for a promising future. Each girl came with a story of her own, but the shared story was the one about how she never expected to be there. It began with the horror of watching those two lines darken and spread on the pregnancy test strip like blood stains on cotton underwear, spots that hadn't appeared as expected. This sometimes occurred in a CVS or Walmart bathroom stall, the only place she had the privacy and opportunity to test herself with the strip she'd plucked from the box on the shelf, then concealed under her shirt. It was only the beginning of the story though, and when she came to us, she had no clue what would come next.

Relationships were often the first casualties. News of who got "knocked up" spread across the gossip-vine at the speed of electrons powering a blow dryer. MHP was geographically disconnected from the main high school campus only a couple of miles away. The distance helped buffer students from the jolt of lost relationships, of friends—even those they'd considered besties—turning a cold shoulder, or worse, turning a friend's predicament into fodder for slut-shaming.

In a recent online conversation, Monica told me she knew a lot of her peers were having unprotected sex, and their judgment infuriated her. Monica was a National Honor Society student and continued attending meetings at the high school across town where she'd been enrolled before getting pregnant. "I get a lot of looks. And they think they're whispering, but I can hear them."

"It may not be what you think," I said. "Some of them may just be terrified of getting pregnant too. You remind them it's possible."

Monica and her classmates were like divergent land masses—tectonic plates—moving in opposite directions. The separation allowed hot, molten rock to rise to the surface and cool, creating a new layer of crust. The girls on either side of that new territory couldn't easily scale the mountains or cliffs; perhaps they couldn't even see each other across the distance. They were no longer on common ground.

***

The teen moms I taught were overwhelmed by anxiety at having children before they were ready to give up their own childhoods. Their bodies experienced sudden and dramatic changes: incessant heartburn and painful ruptures in abdominal muscles that allowed a fetus to grow within its uterine boundaries, creating constant pressure on their moms' bladders. Most girls were terrified of the labor and delivery itself, the pain and aftermath. Would they ever look normal again? Then, once they'd given birth, there was sleeplessness—night after endless night. Colic, diaper rashes, and off-kilter circadian rhythms. At MHP, students found others who shared their fears. Teachers encouraged them to seek support from each other in addition to our campus counselors, nurses, and childcare staff. Girls were reassured that women had borne such discomforts throughout history. They were not alone in their distress. The bonds they formed with other teen moms were built on common ground.

For the girls, prom was one highlight of an otherwise traumatic year. One occasion to just be a young woman without adult responsibility. Everything else we did at MHP supported our students' education, promoted healthy pregnancies, and modeled good parenting. The "Mom Prom," as the girls called it, was a festive party for our 50 or so students and their dates. The Prom Committee and counselors secured a venue and planned elaborate themes and menus months in advance. The teens cast aside their worries for one glorious, fun-filled night. They dressed like princesses and celebrated themselves … those with and those without the baby bump. Many of the girls had dates, some with their baby's father, some with other young men. Some came with a brother or cousin, but many just came with each other.

"It's more fun to go with my friends," students often told me. "Less drama."

"I agree," I'd say. We had way too much Mama Drama already. "My date is Mrs Baker." The social studies teacher and I had a standing date for prom.

Convergent plates force oceanic crust to descend below the earth's surface, melt again, and then harden into a new form: granite. Just like tectonic plates moving toward each other, too much togetherness on a small campus led to occasional earthquakes, and even volcanoes. For the most part, though, girls became fast friends, drawn together by common situations. The relationships that formed between them through the most challenging years of their lives could become rock solid, like granite.

***

The Margaret Hudson Program nonprofit arm, funded primarily through the United Way, managed two campuses, one in Tulsa and one in nearby Broken Arrow, where I taught. One year, we hosted a joint Mom Prom. The campuses served different demographics, a disparity that wasn't obvious to me until they gathered for Mom Prom in April 2014. The gathering reminded me of the dance scene in *West Side Story*, with the Jets on one side of the floor and the Sharks on the other. But there was no Tony and no Maria willing to bridge the gap. Each group danced and interacted in their own style, barely acknowledging the other. The night was charged with equal exuberance and tension.

As partygoers began departing, one of the Tulsa girls couldn't find her phone. "All my baby's pictures are on that phone," she wailed, loud enough to capture the entire ballroom's attention.

Our principal, Genell, quickly assessed the problem, locked down elevators, and identified a suspect: another Tulsa student's date. She confronted him and suggested he empty his pockets, but he refused. There were veiled threats by the young man, whose date demanded an apology for the implicit accusation. Eventually, he complied. Not surprisingly, the phone appeared and was restored to its owner. The young man's date continued to demand an apology from our principal for what she considered a rude method of confrontation. In a bid to restore peace, Genell offered a terse apology, one that challenged any further discussion, and the agitated teens soon left. Her command of the situation reinforced my appreciation for her leadership and our comparatively respectful student body. Their often vocal daydreams of a fairy-tale life with a handsome prince annoyed me, but I preferred their naivete to what I perceived as their counterparts' open hostility. I doubt any of the Tulsa moms saw themselves as princesses. I expect their fantasies included pit bulls rather than princes.

Like girls from vastly different backgrounds, tectonic plates slide past one another at what are called transform, or slip-fault, boundaries. Seismic shifts can arise suddenly, like the panic of a young lady whose entire collection of baby pictures is gone in an instant, or violently like the fiery response of an accused phone thief. Exterior rocks are pulverized as plates grind by each other in opposite directions. Surface layers are dislocated by the encounter. In time, they settle into unfamiliar but relatively stable landscapes, which are more easily disturbed thereafter.

Not long after, public support for programs like ours withered, and the United Way chose to fund other organizations. Grant funding couldn't replace what had been lost, and two years later, the Margaret Hudson Program closed its doors. For more than 50 years, it had fulfilled the needs of a relatively small cross-section of young women, providing support vital to their well-being and the future health of their families and children. Statistically speaking, without intervention, only about half of pregnant teens graduate from high school. We were always aware of the stakes tied to that achievement. A high school diploma is the first line of defense against lifelong poverty, prison, addictions, abuse, and a host of other challenges that might otherwise plague both young mothers and their children.

While I had the privilege of teaching them, the teens in my class-room experienced frequent changes in perspective. Girls clashed with classmates and teachers sometimes, but they also dissolved into each other's arms when a boyfriend cheated or deserted them. They sometimes turned to staff to answer questions they didn't feel safe asking at home. Despite inevitable conflicts, most girls understood that we wanted the best for them and trusted us to answer their questions honestly. We encouraged leader-ship skills through a Leadership Team and an annual Leadership Conference, headlined by inspiring community leaders. The "Day of Caring" was a day set aside each year by the United Way to serve the programs it funded. Participating organizations provided vol-unteers who pampered our girls each year with fun activities and spa treatments. Junior Achievement, a global organization that provides programming for youth in a number of subjects, pre-sented classes in financial literacy and entrepreneurship. Every event we hosted demonstrated our confidence in our students' ability to become good mothers and responsible citizens. Our primary goal for these young women was high school graduation, a rock-solid foundation from which they could build more hopeful futures for themselves and their children. The prom was a chance for teen moms to let loose and celebrate their youth before assuming new roles they'd barely been able to imagine when the school year began.

Students weren't the only ones navigating change, though. Teachers adjusted to individual personalities and situations through all manner of successes and failures. Teens and staff moved together, apart, or side-by-side, challenging each other to do better while accommodating each other's needs. Boundaries were always shifting. I'd begun this later-in-life teaching career with very little teaching experience, thinking I had wisdom about life to teach along with the science I loved. My students and I tolerated each other's inexperience … sometimes gracefully, sometimes not. We all grew as a result. More importantly, meeting students where they were and remaining open to how they could change me led to more trusting relationships.

United States 2020

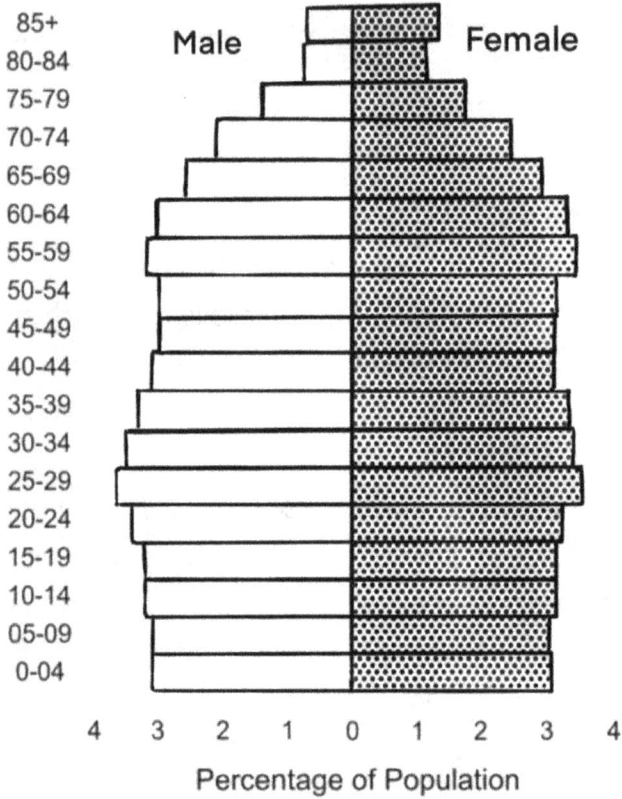

Male   Female

85+	
80-84	
75-79	
70-74	
65-69	
60-64	
55-59	
50-54	
45-49	
40-44	
35-39	
30-34	
25-29	
20-24	
15-19	
10-14	
05-09	
0-04	

4 3 2 1 0 1 2 3 4

Percentage of Population

21
When women do better: We all do better when women do better

"Look at the population data chart for the different age groups." I pointed out the row representing children 15–19 in my environmental science class one morning. "Your age group represents 20 percent of the hypothetical population, and the oldest group of 60 years or older is 10 percent." The other age groups listed in the sample exercise were also multiples of ten to make it easy to calculate. "If this country had a population of 15 million people, how many are in the oldest age group?"

"Can we use the calculator on our phones?"

"I don't think you need it." I wondered if this was just a ploy to get their phones out.

"That's the only calculator I have."

"You can't calculate 10 percent of 15 million in your head?" It was clear I'd underestimated how much math intimidated my students—as it does many adults.

They couldn't, so I relented and let them use their phones. The creation of a sample population pyramid to demonstrate a country's demographics should have been a simple ten-minute affair but ended up taking the whole hour. I was used to my freshman class struggling to make calculations, but the environmental science class included juniors and seniors who'd already taken Algebra. Couldn't they move a decimal point in their heads? It was clear the lab I planned to start the next day using actual countries' population data and statistical calculations might take more time than allotted in my lesson plans for the week.

***

According to the Centers for Disease Control and Prevention (CDC) 2022 National Vital Statistics Report, teen pregnancies have been declining in the United States since the 1990s. The rate peaked at about 6.2 percent in 1991 and has been falling ever since. The CDC suggests that this is so because more teens today are abstaining from sexual activity or are using birth control. This is good news, of course, but the number is not zero. In 2022, the rate had decreased to 1.36 percent of 15- to 19-year-old girls. While MHP enrolled students as young as 13 and 14, the older age group was much larger. Again, the good news is that rates of teen pregnancy have declined by double digits in the past 10–15 years, but there were still nearly 20 million girls aged 15–19 in 2022, based on their percentage of the US population. How many are pregnant? You do the math. Use your phone if you have to. Hint: it's more than 270,000.

Oklahoma, where I taught, had the fifth-highest teen pregnancy rate in the nation in 2022. The rate was 2.12 percent, nearly twice the national average. One of the reasons may be the lack of sex education courses in many Oklahoma schools. It could be because of the inaccessibility of birth control or abortion care in the state. Most of our students took advantage of a Planned Parenthood office a few blocks from campus for contraceptives and spared themselves a second pregnancy before adulthood, which is not uncommon among teen moms. About 15 percent of each year's teen pregnancies are second pregnancies. Our program did a good job of squelching that rate, but we had a girl pregnant with her second child most every year.

Okay, enough with the math.

***

In my environmental science class, our calculations of birth and death rates and the ages at which both occurred helped us determine the environmental and human health effects of those vital statistics. We compared numbers for high-, low-, and middle-income nations, and then compared birth and death rates with educational levels for males and females. The activity came from a chapter in our textbook called "Human Populations," but I supplemented it with an online resource, as I often did, where the lesson was titled "We All Do Better When Women Do Better." Both the text and the online lesson focused on statistical data displayed on charts called "population pyramids." The point of the online lesson was to direct students to the connections between gender, birth rates, and education levels of the countries studied. We concluded that when women are better educated and have access to birth control, birth rates and death rates both decline—factors which correlate more closely with high-income nations. When this occurs, the lesson pointed out, productivity and economic opportunities generally rise, benefiting everyone in the nation.

When we'd completed the activity, our data research supported this conclusion: The well-being of women is key to a country's overall prosperity. A nation's economy is certainly more complicated than what these few pieces of data show. However, to demonstrate population growth and demographics, population pyramids provide an opportunity to make simple graphic connections for teenage girls who are data points on a gender-age-birth rate chart.

My students' experiences corroborated our conclusions time after time. They'd failed to utilize preventive birth control effectively and faced an unplanned pregnancy. Each girl we educated made a choice—or a choice had been made for her—to carry her child to term. The Margaret Hudson Program honored their choices by fostering the completion of their education in a rigorous but flexible academic environment and providing a physical and emotional support system. Staff got to know each girl and her child intimately—we worked with each of them to resolve individual challenges as best we could.

Programs like MHP are virtually nonexistent now, putting approximately 270,000 girls who are facing a teen pregnancy today at risk. If the statistics concerning graduation rate prove true, many of those girls will drop out of school in the next few years because of a pregnancy, leaving them and their current and future children vulnerable to many preventable failures. The reason this situation exists? A well-designed program for teen moms is expensive.

Unfortunately, we tend to be a short-sighted society when it comes to where we spend tax dollars. We think only of the cost of educating those relatively few girls but ignore the long-term costs to all of us if they drop out of school. Many of them start in precarious situations in terms of financial assets. Without economic resources, every obstacle looms larger for these youth, who often develop health challenges as well as drug or alcohol addiction. Such addictions can destroy entire families.

Society pays monetarily if the child is unemployable and depends on public assistance. In many cases, these young moms or their children end up in prison, where the public decries the paltry expense of their upkeep without regard for the lost social and intellectual capital while they're incarcerated. Mothers who are imprisoned cannot support their families, either as caregivers (their expected role) or as wage earners (usually a necessity). These losses extend beyond incarceration as well. A criminal record is often a death knell when it comes to rising above poverty—ex-convicts in some states cannot vote, utilize public programs, or enter certain employment fields.

Our democratic government suffers when citizens cannot think critically or participate intelligently in choosing leaders and policies that benefit all. Less educated members of society are more likely to be used to someone else's advantage, depriving them of autonomy. Programs like MHP attempted to instill logical thinking strategies that allowed students to discern the most beneficial choices for themselves and their children. When such programs closed, pregnant teens were shuttled to alternative school campuses if they were available; not all had that option.

Many instead chose to remain at their local high school. In either case, the support designed specifically for teen moms and their children is rarely an option. Many of these girls will not succeed, and what they have to offer the rest of us will be lost.

These outcomes are symptomatic of a country that doesn't acknowledge the value of women or the truth of the statistics my students clearly illustrated with their demographic calculations. Our nation would do well to pay attention to the lessons my students learned.

We all do better when women do better.

22

Eight years: Humility and empathy are a teacher's most powerful assets

"Isn't she adorable?"

"Absolutely. She takes after her mom." The photo my student offered me captured the gleaming eyes and open-mouthed grin her daughter flashed at the exact moment before her lunge toward the camera. "I love your matching shirts, too."

Another girl might say, when handing me a photo of her grinning toddler, "He's got his daddy's dimples," eager to connect this child with the boy who'd stolen her heart, at least once and for a season—the dad who may or may not acknowledge this child as his.

"He's very handsome." I'd answer with a smile. "Did you write your names on the back?" With so many new students each year, I'd learned not to trust my memory from one year to the next.

Some pictures were of just moms or just babies, but usually, the girls handed me their "Mommy and Me" yearbook photos taken a few weeks before, when the district's photographer came to campus for our annual Picture Day. Wailing and tears emanated from the cafeteria off and on all day, punctuated by the frantic sounds of the photographer oinking like a pig or squeaking the rubber clown he danced in front of the children. Shrieks of delight were drowned out by the howling. When students pressed the photos into my hands later for my memory board, the stress of wrangling an infant into the perfect, brand-new blue and green astronaut onesie or toddler into her gauzy pink princess tutu without causing a meltdown was forgotten. The acrobatics required to get the child to smile at the right moment was but a blip of memory.

I pinned the photos to a foam core board with a straight pin through the slim white space that surrounded the faces set against a generic backdrop of the blue clouded sky or what appeared to be a floating castle in a pinkish sky. After teaching for eight years at MHP, I'd collected dozens of photos of students and their babies. Those shining faces greeted me each morning when I entered the classroom and gave me a reason to smile. When I left the classroom for good in 2015, those same photos brought tears to my eyes.

A lot can happen in eight years. You could go to college or technical school, get advanced degrees, and maybe add some initials before or after your name. You could have a couple of kids and settle into a career, perhaps blow through a couple of jobs. You might lose someone you love or meet the person of your dreams, that once-in-a-lifetime someone. In eight years, you could run a marathon in every state in the United States. If you're lucky, you could win the lottery and move to Rio de Janeiro, if that's your kind of thing. There's really no limit to what you could do in eight years. In eight years, you could even learn what it means to be a teacher.

***

I bought a black plastic bin on wheels at Office Depot the week I started teaching high school. The bin hauled my blue spiral attendance book, a 2 inch binder with tabs and lesson plans for the four different science subjects I taught, along with 10 lbs textbooks for each of them, and an indestructible, black-handled vinyl tote bag decorated with black spirals on a white background. Inside the bag were daily handout masters, markers, and assorted supplies I carefully assembled the evening before. Most days, there were disposable lab supplies inside: a roll of paper towels, a lemon, a half-full bottle of vinegar, maybe a box of straws, or a carton of live crickets to feed the gerbils. In eight years, the rolling cart had transported thousands of household chemicals, foodstuffs, desk-drawer odds and ends, and hand-held tools from my husband's tool chest to my classroom. Wedged among them were my frustrations, as well as my hopes, dreams, and fears for my students.

In 2015, the bin was rolling in reverse. It carted home the paraphernalia I'd bought and left in cabinets, on shelves, and in the large supply closet in the back of the classroom. Every spool of ribbon, bottle of clear Elmer's glue, half-empty box of food coloring droppers, or alligator clip had been vital to its lab or demonstration in one class on a particular day. But now, the flotsam and jetsam were going home with me. Part of me feared being accused of hoarding useless stuff by my successor, which current evidence on overstuffed shelves might support. I just never knew when I'd need that 6 inch wide strip of green cellophane. Sure, the biology plant lab that used it hadn't worked three years in a row, but I was married to my hope.

After a couple of weeks of reverse-stockpiling, the lab storage shelves were looking emptier—just as they'd appeared when I started. An entire long wall was filled with cabinets and an impermeable black resin lab-grade countertop. In 2007, the cabinets contained nothing but stacks of textbooks. A plastic human-sized, headless dissectible torso I'd obtained from a grant now perched on the counter between the two sinks, one of its interchangeable sex organs neatly stowed in the drawer below it. The bio plant lab and shelf took up a large section of the countertop near the door just as it had when I arrived.

The two floor-to-ceiling bookshelves along the back wall were now about half full, similar to their original state. I lugged home the books I'd brought about amazing science facts, reference books, three years of *Science* magazines, and a few classroom sets of novels I'd bought for my junior English students to read and appreciate. I emptied the files from my desk drawers and pulled more reference books off the shelf behind my desk to drag home too. Despite the dent they made in my monthly paycheck, my eternal and outsized hopes for their effectiveness in engaging students outweighed any bitterness at what they cost me personally.

The last thing I removed was the memory board full of smiling students and baby photos. It was crudely constructed from a large, square scrap of foam core my husband brought home from his office. I bought a cheap black knit remnant from the fabric store and stretched it over the foam, stapling the edges over the backside. A couple of the other longtime teachers had similar boards in their rooms when I started teaching, but it didn't occur to me for a couple of years to replicate the custom in my room.

In time, though, girls began pressing photos into my hand. "Here, Mrs Airhart. Isn't Bailee's outfit precious?" they'd say. Or "Damon was crying until the photographer played peek-a-boo with him, and then he wouldn't stop laughing. You can see his two top teeth when he laughs. Isn't that cute?" Of course, I agreed. Sometimes the girls gave me photos of just themselves, "to remember me by." Before long, I started asking the girls for photos. By 2015, the board was covered.

During the last week in May, I gingerly unpinned each photo from the board and ensured there were names on the backs. Names were sacred to my students, and therefore to me. I was deliberate about calling every baby by name. If my recall wasn't quick enough to supply it on occasion, they were simply "Sweet Pea."

As I placed the photos in the greeting card box I brought to store them in, I wondered what would become of the girls and their babies. What had already become of the ones from the past years? I feared the worst for some and held onto high hopes for others. Letting go of my fears and hopes was challenging; letting them go out into the world was hard all by itself. The toughest lesson they'd taught me was to trust them to be who they wanted to be. Even when my aspirations for my students were no longer relevant to them, they were relevant to me. I couldn't part with them; they went home in the cart too.

A final scan around the classroom before I turned off the light and locked the door for the last time revealed a science class-room like any other. There were no obvious signs of the special population it served or of the teacher who'd spent eight years with them here. I thought briefly of the 55-year-old self I'd been in 2007, how fearlessly she'd tackled this radical twist in her career journey, and how wrong she'd been in imagining the course it would take.

I'd taken this job hoping to improve the futures of girls who were faced with a life-changing circumstance I thought I understood. To a certain extent, I'd accomplished that, but not in the way I'd expected. I'd imagined filling their minds with the science facts I loved, and in the process, imparting valuable insights about life and motherhood. I'd experienced a good bit more of both, after all. Unfortunately, none of my students had the same life I'd had. My insights didn't often apply. Stories about my experiences didn't impress them. There were days, though, in which the wisdom I shared found a necessary target. The frequency of those moments accelerated over time and kept me going.

At the outset, I didn't understand as much about my students as I thought I did. They weren't like each other, and they weren't like me as a teen. They weren't even like me as a pregnant teen. Education had also changed a lot since I'd been a student, and I had to adapt to new ways of communicating. My students had a great deal to teach me about filling their needs. But before I could become the teacher I'd imagined I would be in 2007, I had to learn to approach students with humility and empathy. I've come to believe these are a teacher's most powerful assets.

A lot can change in eight years.

Part III
2015–2024

Grow

In any given moment, we have two options: to step forward into growth, or step backward into safety.

—Abraham Maslow

Part III Introduction: Leaving the high school classroom

In 2015, after spending eight years in the classroom with pregnant and parenting teens, I realized that my worldview had changed. While I thought I understood the population I would be teaching, I had underestimated the complexity of their lives and, over time, learned to respect their goals and priorities. I made many mistakes along the way but stumbled upon successful teaching strategies too. Teaching instilled in me the ability to approach students (all people, really) with humility and genuine curiosity, and this transformed me.

Some of what I learned about teaching is disturbing, however. In recent years, trust in educational institutions has eroded. Schools have a unique responsibility and opportunity to shape students' futures and, by extension, our society. Appropriate investments of public support and funding sources are critical to creating an atmosphere that encourages students to become responsible citizens. I have deep concerns about current attitudes toward public education, especially where pregnant teens are concerned.

Teachers seldom know how they impact their students' lives, and the reverse is also true. With the advent of social media, which was a scourge in the classroom, I am now able to follow some of my former students and their successes. While their successes are all their own, I am grateful for the opportunity to be a part of their journeys. I celebrate with them and grieve with them today. Most of all, I'm thankful for all my students taught me while we were together.

Lessons from teen moms

- Failure is an excellent teacher.
- Students can be teachers too.
- Take risks but know your limits.
- Schools provide valuable support systems for students.
- Prepare students to move independently into the world.
- Learn correct names, then use them.
- Find what works and repeat as necessary.
- Preview all class materials thoroughly before presenting them to students.
- Lessons shaped by passions or convictions often have greater emotional impact.
- Honest concern for student welfare is the key to their academic success.
- Don't place too much faith in structures alone; technology is a means to an end.
- Recognize biases and work to overcome them.
- Don't underestimate students' abilities.
- All students can achieve if they're given enough encouragement.
- Meeting students where they are promotes trust.
- We all do better when women do better.
- Humility and empathy are a teacher's most powerful assets.

23
Philosophically speaking: An evolving view of educational philosophy

"It's your turn," Holly, the Family and Consumer Science teacher, announced with a snicker. This was after the impromptu all-campus assembly one October morning in 2009. Our principal, Genell, had gone back across the hall to her office, and the girls were filtering back to their classrooms.

I'd only been at MHP a couple of years and still existed in a fantasy bubble of smug confidence as a Good Samaritan, changing the lives of teen moms. The bubble had thinned, though, and I'd begun to question my ability to pull it off. Teaching doesn't pay well, and other rewards, like student appreciation or community respect, are hard to quantify. Being chosen campus Teacher of the Year (TOY) meant I would also become a nominee for the district Teacher of the Year (TOY) award, and I was pleased. At least my *colleagues* respect me, I thought.

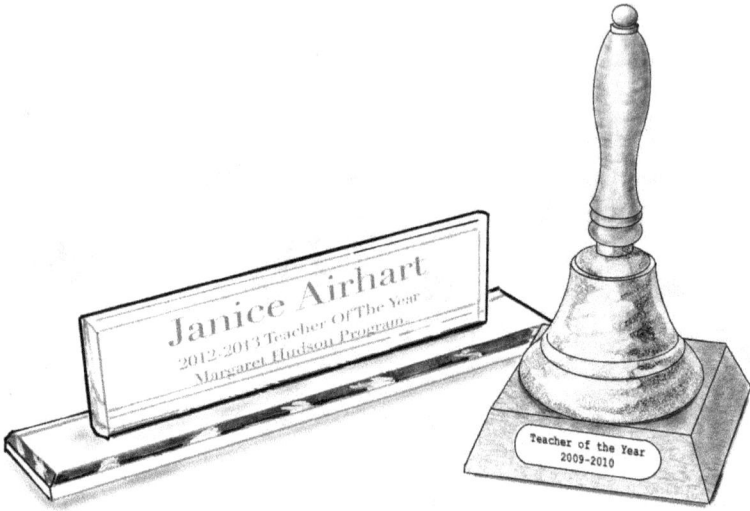

From Holly's grin, I could tell she saw being named site TOY as more of a load of work than an honor. "I hate the whole thing," she said. "The portfolio writing, the interviews. All of it's a pain in the ass."

I nodded halfheartedly. "It does seem complicated."

"They'll *never* give it to one of us," she said, referring to the competitive district-wide nomination with a half-smile, her hand resting gently on my arm holding the lovely arrangement of fall flowers I'd been presented with a few minutes earlier. I was well aware of our program's standing in the district, something like a red-headed stepchild.

"Congratulations," Linda said, without contradicting Holly's assessment. "If it'll help, I'll send you my portfolio from last year as an example." Despite what she may have thought about Holly's comment, as my teaching mentor, she always supported me.

"Thanks. That would be nice." I *enjoy* writing, so composing the portfolio didn't sound like such a hassle. That was before I saw the guidelines and read Linda's polished example. For a few days, at least, I enjoyed the flowers and everyone's good wishes.

***

The district required that each school site nominate a TOY in the fall. Since there were so few teachers at MHP, Holly was right about taking turns. Even though all the teachers were well-qualified for the district honor, it seemed more that selection as the MHP site nominee was just a matter of taking turns—even part-timers like me. Holly and Linda had been teaching for many years before I came, so they'd been nominated multiple times. It wasn't until I was nominated again in 2012, by then teaching full time, that I understood Holly's cynicism. My turn wouldn't have come around quite so soon if Linda had not resigned in 2011 and the pool of experienced teachers shrank (I'm taking liberties by including myself in this number). When Holly resigned in 2013, I became the MHP teacher with the longest service record in the district—a whopping five years. I'd already been the oldest teacher on campus. When I was the site TOY nominee again in October 2012, what had been clear to Holly and to Linda was finally clear to me: They'd *never* select one of us for District Teacher of the Year.

***

As Holly indicated, completing the portfolio was indeed a tedious process. Once the warm and fuzzy feeling surrounding the nomination had subsided, it became clear that meeting the deadline for submitting my portfolio would require speedy but extensive reflection and adept writing skills. Portfolios didn't have names on them, so each submission was anonymous until after the finalists were selected. Once finalists were chosen, the district TOY committee would conduct interviews and a crew would come to video a class session. Those were stress-inducing to my deeply introverted nature, but I didn't give them any thought at the outset.

I went to work on the easy parts of the portfolio first, filling out the form with my work history and basic facts about my education, professional memberships, and previous awards. All of it was perfunctory. I wanted to take my time with the narrative sections: Professional Biography, Philosophy of Teaching, Education Issues and Trends, Teaching Profession, Community Involvement, and Success Story. Each of the longer sections had a two-page limit. A couple of other short sections rounded out what would become a fifteen-page portfolio.

For the longer sections, I toyed with ideas and outlined for days before composing what I considered a polished document, consulting Linda's example portfolio as I went. I think I asked her to review my draft and make suggestions before I finally edited and submitted it.

A few weeks after my submission, someone from the District TOY Committee paid me a visit between classes to inform me I'd been selected as a finalist for District TOY.

"Yours is one of the best portfolios I think I've read," she said. I've always wondered if I'd have been eliminated earlier if they'd known which portfolio was mine.

"Thank you." I had labored endlessly over the text, revising and tweaking until it expressed exactly what I wanted to say.

"But," she said. "since you're part time, we won't expect you to complete the interview and observation phase. I'm sure that will be a relief." She smiled warmly and patted my arm briefly.

"Okay." I wasn't sure what else to say. She was right that it was a relief to be done with the process, but it felt like a letdown.

"Congratulations, though."

When I told Genell later, she was livid. "Oh, *hell* no!" (I exaggerate. Genell would never have said hell.)

"You deserve to have as much opportunity as the other candidates," she said, shaking her head. "I'll call them. You will be included in every step of this process." Genell was resolute.

I appreciated that Genell felt I deserved an equal shot, but if she got her way, I'd have to complete the committee interview—interrogation—and someone from the technology department would record me in the classroom delivering a lesson. The committee would then dissect my every move … and misstep. Besides, the obligations of the TOY were enormous, representing the district at events that would stretch my capacity for schmoozing beyond my perceived limits. I didn't really want the responsibility.

"I don't know," I said, but she just shook her head again.

Genell prevailed.

In the following couple of months, I think I did reasonably well on the next steps, and I invited the other teachers to attend the TOY awards ceremony to join our school's table for dinner. We got the typical baked chicken with mashed potatoes and seasonal vegetable blend, and I got to stand on stage with the other finalists, while the District Teacher of the Year was named. It wasn't me, and I felt a mix of relief and disappointment. But I brought home an attractive brass bell with "Teacher of the Year" and the school year engraved on the base. I had a new award to put on my resume.

***

When I was voted site TOY again in 2012, I was full time. Despite all evidence to the contrary, I remained optimistic. Surely, the main reason I fell short of the district honor in 2009 was because I was part time, I thought. Part of me understood that Holly's prediction of my chances—the chances for any of us—was accurate.

Yet hope dies a hard death. I honestly didn't realize I had so much hope in me.

At this point, I was serving on an Oklahoma State Standards committee for biology, working with other biology teachers across the state to write academic standards for use in all Oklahoma public schools. Every month or so, I went to the state capital in Oklahoma City to help draft the language that the legislature would vote to adopt—or not—before the year was out. I'd also previously served on a statewide Science Leadership Team with other Oklahoma science educators. Perhaps being chosen for these esteemed groups would give me an advantage.

I dusted off my 2009 portfolio, making updates and revisions. Because I'd entered a sphere of suspicion about the likelihood of taking home the biggest honor, I didn't devote the time to my portfolio this time around. Looking back, I can understand that the district wanted a candidate to represent them who had a role more representative of a mainstream school in the district. MHP was *not* that school. But I put in new material, new trends, new success stories, and an updated philosophy of teaching.

Site nominees in 2012 were required to ask a student to escort them onto stage at the celebration. I asked Irina, who'd been in at least one of my classes for three years. She shouldn't have been able to stay at MHP that long because the official maximum limit was two years. But Irina, who'd immigrated from Russia with her mother and much younger brother a few years before, stubbornly announced her intention to drop out of school if we made her leave, and Genell let her stay a third year. She was due to graduate in a few weeks.

When I asked her one morning before class if she'd escort me onstage at the banquet, she made me repeat my question. "Really?" she said, a little surprised, but beaming. "Yes! I'd love to!" She couldn't stop smiling after that and kept looking up at me through environmental science class as though we had a secret. I'm pretty sure she winked at least once.

At the Star Gala event in April, where the district TOY was named, Irina escorted me on stage. The evening's emcee started and stopped a couple of times to introduce us, and in the end mispronounced Irina's last name anyway.

"Mrs Airhart teaches four different science classes, plus Grade 11 English, creative writing, and journalism," she said. "Wow. That's a lot of classes."

After all the site candidates were introduced, there was a bit of tension while the emcee announced the names of the finalists, none of them mine. Those of us who weren't chosen went back to our seats. At our table, Irina proudly passed around pictures of her daughter to the friends I'd invited. I was disappointed, but not much. At least I'd gotten a nice chicken dinner and a heavy, acrylic desk ornament with the year engraved on it. I suspect Genell was more disappointed than I was.

***

The best thing to come about after being nominated twice for District Teacher of the Year was the requirement to write a portfolio. It forced me to think about my philosophies around teaching, education in general, and what constitutes success. Most of the time, I was too bogged down in getting next week's lesson plans complete to think about pie-in-the-sky concepts. Sometimes, even tomorrow's plans were a frantic scramble. By 2012, I'd learned quite a bit, and I'd had enough time to think about what I was doing so I could write something coherent for the TOY Committee. Whether I was considered a fit candidate for the district honor or not, I'd finally reached a point where I knew what I was doing and why I was doing it … at least most of the time. It had little to do with the complex bundle of good intentions I'd crammed into my book bag when I first stepped into the classroom.

I thought I had the wisdom to share with a specialized population of girls, and while I did have some, I learned I didn't know as much as I thought I did. In the same ways that my own children taught me how to parent, my students taught me how to teach. I'd learned how to better recognize their needs and some strategies to meet them when I could. My deepened empathy and compassion compelled me to try my best.

I thought I needed to prepare my students for the world as I saw it, for the abilities I'd needed when I was their age—back in what they considered the dark ages. Preparing them for college and a career in the twenty-first century required different skills and greater patience than I'd counted on. Some things had not changed, however.

In the "Professional Biography" section of my TOY portfolio, I stated, "I'm driven by curiosity … It is the most valuable characteristic I hope to develop in my students as well; it is vital to producing self-directed, life-long learners." Students were far more curious about some topics than others, and most of those topics had nothing to do with science. I had to fit course objectives to their unique curiosities in a way that encouraged learning.

I wrote about my desire to remain a life-long learner again in the "Philosophy of Teaching" section and mentioned ways I'd tried to channel student curiosity to science by becoming creative in planning lessons. It's been said that the best way to learn a subject is to teach it, and I proved the truth of that concept. "One of the greatest rewards of teaching is the opportunity to learn something new, and to transform that knowledge into a learning experience for my students—not just for a good grade or to pass an exam, but because learning is its own reward." In 2012, while I was rewriting this section of the portfolio, I also maintained a blog titled "Learning Is the Reward," with a tagline of "No experience is wasted, unless you fail to learn from it." It was then—and is still—my personal motto.

In another section of the portfolio, I identified some disappointing issues and trends in education that continue today. There has been a growing, almost alarming, lack of trust in educators and education as an institution in recent decades. While I suggested some hopeful remedies nearly a dozen years ago, I'm not as certain today that the trend is reversible. One deeply held belief I expressed still holds true: "I am absolutely convinced that equal access to public education is essential to a democratic society." Looking back at my years as a high school teacher, from 2007 to 2015, I see a slow but continued erosion in the support for institutions across the board: education, law enforcement, healthcare, religious, social welfare, and many more.

It seems disagreement about the mission of public education in our society has led to hostility and rancor, which leads to one essential question: Who has the right to determine what a child learns? Perhaps the parents or the students themselves should determine what constitutes an adequate education for lifelong success. Perhaps professional educators or district administrators have enough experience with teaching strategies and their outcomes to create a framework for a thorough education. Perhaps society has a right to determine what a child will need to participate more fully in determining a future that functions well for all people. If the latter is true, there may be a role for legislators in enacting boundaries that support that future. However, in my observation, most legislators do not have the training, experience, or sincere desire to appreciate a "good education," and would do best to rely on others to define them. Instead, what I see is legislation designed to curry favor with the parents who elect the lawmakers. While parents are most likely to have their children's best interests at heart, I'm not sure they know what their children will need to succeed in life, and within the society of the future. Whose vision of success is valid anyway? Each of these groups has a stake in what a student learns, but when they can't agree, forward motion is stymied, and all suffer ill consequences. This is where public education finds itself today.

As for me, writing the TOY portfolio in 2009 and again in 2012 forced me to define success as a teacher for myself. I'd embarked on this career for several wrong-headed reasons. Teachers were in demand both then and now. The demand is so great that schools can't afford a bias against older workers. I'd had wonderful experiences as a student myself, and I thought I could duplicate them for students, not realizing the truth of how much had changed in the interim. I soon learned that many—though not all—of my academic experiences were irrelevant, maybe even obsolete.

During one summer between school years, I remember spending a good deal of time studying Carol Dweck's theory of growth mindset. She contended that each of us, and students in particular, have either a fixed mindset (a person is born with a fixed intelligence level) or a growth mindset (a person can grow in intelligence through practice or hard work). In the following semesters, I ensured messages to my students communicated my confidence in their ability to succeed if they put in the time and effort. I could be the poster child for this particular message, after all. I certainly knew nothing about how to teach in 2007 but I had eight years to practice. I'd never worked harder at learning anything in my life.

I was originally motivated by a desire to *save* these girls from the dangers of poverty and its associated effects, believing that without intervention, too many would fall prey to those ills. And I believed I was qualified to intervene. I was sorely mistaken, as I've come to realize after observing one student after the other create successes on her own terms, success that didn't fit how I'd once characterized it. In time, I learned to respect *their* definitions of success, no matter what situations or individuals lent it meaning. Many of my students had growth mindsets too.

Holly was right about my chances, the chances for any of us at MHP, of succeeding at being named District Teacher of the Year. The designation comes with a lot of responsibility and notoriety. It's certainly an honor. However, there are other rewards for a teacher that I've realized with the perspective of time. Whenever a student trusted me with a picture of her baby, with a story of her family's struggle as undocumented immigrants, or asked me for help with a personal dilemma or a difficult lesson from her vocational-technical certification course, I was honored. Being honored by my colleagues or my district was nice, but it wasn't the honor I'd set out to achieve.

24
Getting social: Following students beyond the classroom

"Please put your phone away." This refrain was issued in class nearly every hour of every day, sometimes multiple times per hour. Some students denied they were texting or scrolling, but a good teacher, just like a good parent, has a pretty accurate BS meter.

"I'm not …"

"*Put it down.*" Sometimes, I'd cock my eyebrows and point to the sign on the door that announced, "If I see it or hear it, I take it." Aside from the seeing and the hearing, there were other ways to detect deception.

Telltale signs: (1) a student seeming suddenly fascinated by her crotch, (2) a sudden grin or chuckle while contemplating said crotch, (3) hands that stayed in her lap too long, and (4) random, furtive glances up and to both sides before resuming contemplation.

Facebook had only been around a few years when I started teaching high school in 2007, and while some students also used Myspace or Tumblr for a while, there was no denying the Facebook juggernaut. A student scrolling Facebook (in class, no less—however did they manage it?) was like the fly circling my dinner plate. No matter how often I swatted at it, it kept buzzing back. Ubiquitous social media use among teens seemed only a vehicle for shaming, bullying, and mind rotting. It was one more distraction on a multi-service campus full of built-in distractions such as constant calls to the classroom from childcare or nursing, breastfeeding in class, frequent absences, and more.

The school district prohibited staff from interacting individually with students via phone, text, or online platforms. Genell made it clear we were not to "friend" students on social media. As a result, I maintained a professional distance that might have come across as disinterest, although that was far from true. It was important to me to avoid an appearance of favoring one student over another. I presented what I hoped was a uniformly friendly, empathetic, and goal-oriented attitude toward all students. I have no idea if I succeeded. On the inside, I was often torn by competing instincts to throttle or hug, saddened by news of a student's ongoing personal trauma, or overjoyed by the sudden spark of understanding in a girl's eyes. The up-and-down swings in emotion were draining, but I'd spent decades perfecting a calm exterior. It was my superpower.

Teachers anguished over the ineffectiveness of the district cell phone policy, and we tried to structure our own. For a while, we were allowed to take cell phones from students and bring them to Genell's office for students to retrieve later. There must have been an incident in which a phone was lost or stolen on the way to the principal's office, because suddenly, we were told we couldn't take them anymore. Teachers proposed putting a basket on our desks where students could deposit phones until class ended. No, that wouldn't do. We suggested asking students to place their phones out on their desks where we could always see them. That got shot down, too. Surprisingly, students often texted with family members during class. Perhaps Genell was reamed out by a parent who felt the need to communicate with their student whenever the impulse struck, in class or not. I'd never have dreamed that an avalanche of school shootings in future years might make it critically important for students to be able to call 911 from the classroom.

We were stuck with surreptitious Facebook scrollers and posters. One student was deeply admired by the rest of the students for her mastery of concealment. They dubbed her "cell phone girl," mystified that she got away with phone use all day while the rest got caught. I agree, she was good. In my A&P class, I sometimes suspected her, but she maintained a solid "A" average, so I gave her the benefit of the doubt.

Social media has moved on in recent years, introducing applications and platforms at a frequency that rivals the rate of iPhone updates. Their demographics are constantly in flux. Almost as soon as an adult population discovers a new platform and begins invading their children's spaces or feeds, the younger generation empties it out. I wasn't allowed to friend *current* students, but over the years, I'd developed a sizable network of *former* students.

Teachers don't often know what becomes of students, and it haunts our memories. It's hard to let go of our hopes and fears for them. Although Facebook use was a massive headache to me in the classroom, it has become a surprising blessing to me all these years later. Not all my former students utilize this platform, but many still do.

Most former students, who I now affectionately consider "my girls," seem wedded to Facebook. Perhaps it's just the platform they know best. I've lost track of those who've left it behind. But for dozens of others, I regularly click reaction icons, send birthday wishes, or make short comments on their posts. They're all still young women with growing families they're understandably proud of, and most of them have deeply rewarding lives.

From the "Mommy and Me" photos my girls pressed into my hands after Picture Day to the yearbooks chock full of photos that sit on my shelf today, I love physical reminders of this one girl, that one baby, and the stories about each one. Full color spreads depict not only students and babies but also annual holiday celebrations, assemblies, costume and door decoration contests, field trips, and candid shots around campus. Those are lovely memories, but they're static. Facebook allows more real-time interactions and has the power to touch my heart in new ways. Many girls are now married and well established in their chosen careers. A few have made surprising choices, like one girl who insisted she wanted to become a welder (a choice I applauded), but who is now a child protective worker for the state of Oklahoma. I revel in discovering how their adventures unfold.

Not all the news I read about my former students is good news, though. Brooke's mother passed away suddenly after a heart attack. Camille finally found the man who made her feel whole after suffering childhood sexual abuse at the hands of a relative and an adolescence marred by depression. Then her one and only committed suicide. Charlotte, now with two daughters, has married and divorced the same woman twice. Elena's mother, sister, and nephew perished in a devastating house fire. Amanda's third child was born with a rare and usually fatal disease. More than one has been convicted of a felony. As I write this, at least two are in prison, which proves the statistics about the consequences of teen pregnancy that MHP sought to mitigate. These girls, *my* girls, hold a permanent place in my heart. I grieve with them and rejoice with them.

It hasn't escaped my notice that my students were forced to change the course of their lives while I was deliberately doing the same. For these young women, their detours from an imagined future originated in unplanned pregnancies and presented dilemmas their youth and inexperience were unprepared for. By contrast, I *chose* to deviate from a well-worn path of prior experiences, with an exaggerated sense of accumulated wisdom. In a way, the girls and I met in the middle. I shared what I could, but I was surprised to discover that we all had much to gain. The rewards I've realized are so much richer than those I imagined at the outset. I know online profiles don't always parallel reality, and I don't presume posts tell the whole story.

One of the things my students learned was how to persevere through unexpected changes, and I think it made them stronger as a result. They are fiercely protective of their families today. Monica, who was featured in a 2012 newspaper article that resulted in some vicious public feedback, felt she had something to prove to those who predicted her pregnancy would be her downfall. She's now a college graduate teaching biology to ninth graders in the same district she graduated from. Similarly, many of my former students have been fueled to achieve more because of the challenges they've faced. We've all learned perseverance. Teaching young women who are struggling to find a new place for themselves is hard, harder than anything I've ever done. But we learned from each other how to navigate change, and I'm thankful for the lessons.

25
Subject to change: Maintain curiosity and an open mind

What was I thinking in 2007 when I made up my mind to obtain a secondary alternative teaching certificate? I'd hoped to contribute to positive changes in my students' lives. What I hadn't anticipated was how much the experience would change me. My students were patient teachers, though. I'm still processing the lessons they taught and grateful for their tolerance of a third career, sometimes cranky, greenhorn teacher who, for many, was older than their grandmothers.

Change can be hard, but it doesn't have to be. The hardest changes are those forced on us by circumstances out of our control. When we're the orchestrators of our own change, it can be stimulating. My husband and I have moved ten times (so far) during our marriage—sometimes due to his job change and sometimes motivated by nothing more than a desire to change our environment. I embraced each of those changes as adventures. New environments breed new perspectives, and there's nothing I love more than learning something new about a place, a vocation, another person, or a philosophy. Current predictions

are that young adults today will change jobs and careers multiple times throughout their lives, driven by economics, culture, or personal preferences. For those who resist, the road ahead will be bumpy. However you choose to look at it, change is not optional. I decided early to seek out my own changes, before they found me, and often from simple curiosity.

The origin of the phrase "curiosity killed the cat" is uncertain, though a version of it seems to have first been used by Ben Jonson in an obscure 1598 play. His version is "Care'll kill the cat"—care in this case means concern or worry. It was repeated in Shakespeare's *Much Ado About Nothing* the following year. Whatever its origins, I hate the statement. It's used to shut down children's exploration of their world, sometimes as a threat.

Curiosity has led me to some amazing learning opportunities, and I consider my teaching career one of them. Facing a classroom of teenage students should have given me greater pause, but I was curious to know what it would be like to communicate ideas more directly than I could as a writer. In 2007, when I chose a new path full of career potholes, I managed to navigate them in a satisfying way over time. I learned to think on my feet more easily, which has served me well in the time since I left the classroom. I couldn't always dodge the hazards, but I wouldn't trade the bumps for anything now. I'm still on the road, dodging obstacles and encountering new opportunities through writing and publishing books, articles, and personal essays.

All of these recent ventures began with curiosity to know and understand more and to expand my perspectives. Who knows what change I might pursue next? I hope I'm not done yet. A common retort to the familiar warning about feline curiosity sprang up in the early twentieth century, first noted in a 1912 newspaper article from Titusville, Pennsylvania. In response to the familiar phrase, one reporter wrote, "Curiosity killed the cat, but satisfaction brought it back." I love this response. If curiosity kills this cat someday, I will at least have lived a more satisfying life because I let my curiosity lead me down such interesting roads. I may not have a cat's nine lives, but I hope to cram as much learning as possible into this one.

***

My motives for becoming a science teacher seem somewhat foolish now, naive at best. I craved a new source of intellectual and emotional stimulation … and a paycheck. In my mid-fifties, I found the employment search more complicated than when I was younger for many reasons—not the least of them the bias against older workers. Teachers were in high demand, and my dreams of a successful writing career hadn't materialized. Teaching, like writing, transfers information; it seemed a suitable alternative. I'd spent several years volunteering as a mentor for youth in state welfare programs and teaching youth at my church. I also spent a year with the Kids Count program, a project of the Annie E. Casey Foundation, advocating for school-age children and creating a writing curriculum for fourth graders as a result. I delight in the originality of young people. When I was offered a job teaching science where I'd mentored a couple of years earlier, it seemed a divine call to share what I considered extensive, hard-won wisdom. Altruistic? I like to think so. Arrogant? Maybe.

I'm not sure I'm qualified to judge. All I know is how meaningful the changes they introduced have been.

I'm proud of my former students for embracing uninvited changes to their lives as well. I'm proud of every pregnant teenager who finds her way out of the quicksand and into adulthood, despite dire predictions. Many make use of those difficult years as proving grounds. They're often better mothers for what they've been through. They teach their children by example what it takes to succeed, just as they taught me how to become the teacher they needed. I didn't become a science teacher for *all* students, however. My short stint at North, in the overfilled classroom of typical teenagers, proved that. Failure can be instructive, too. Instead of simply becoming a high school teacher, I learned to teach these *particular* students, teen moms, during a challenging period in their lives, while simultaneously challenging myself to grow in mine. Each experience taught me about extending myself beyond what I thought possible.

Since leaving the classroom, I've continued to be involved in the lives of young people, though more recently with much younger children. Before the COVID-19 pandemic, I volunteered with my county's Child Advocacy Center, which offers support services to families and children who have experienced abuse or neglect. Most abuse victims are referred to the center by law enforcement. I now coordinate a Literacy Partner group through Education Connection, an organization in Austin, Texas, that provides reading partners for elementary school students who are struggling to read at grade level. Reading with a child for a few minutes every week gives me great joy. It's a simple thing, but it means so much to a child who needs another caring adult in their life. I also plan to meet soon with a counselor at an area alternative school about mentoring one of their pregnant or parenting students. My life was similarly enriched by such relationships, and I know how important this small thing is.

For the many children and youth I've been able to impact through mentoring, teaching, writing, reading, and lecturing in recent years, I owe a debt of gratitude to a small group of young women who struggled alongside me in their quest to overcome fear and become the women and mothers they wanted to be. They changed me, though not with deliberate intent. They taught me what it means to care for each other during the tough times and how to respond to the needs of others with humility.

I'm so grateful for the lessons the girls taught me. I hope they know I always wanted the best for them, even if neither of us really understood what that was.

Recommended projects and discussions

1. Interview a current or former teen mom about the obstacles to completing her education. What obstacles does she still face? What difficulties, if any, will/did her child encounter as a result?

2. Research the requirements in your area for becoming alternatively certified to teach at the secondary level. What is the ratio of traditionally certified versus alternatively certified teachers? What advantages or disadvantages do alternatively certified teachers provide to their students?

3. Examine your relationship with a favorite teacher or faculty member, and the reasons it is memorable for you. What characteristics or attitudes of the teacher contributed to your positive relationship? How did the strength of this relationship affect your learning?

4. Discuss a step-by-step process for structuring daily lesson plans appropriate for teen moms. What factors would need consideration? How might the structure change if the population were non-English-speaking students? Students in a residential juvenile facility? Urban versus rural students? Students with disabilities?

5.   Choose one or more of the lessons which appear in Part II of the book which you feel are applicable to a special needs classroom. First, define the unique needs of the population you select, then describe how you would apply the lessons in the context of that population.

References

American Society for Positive Care of Children. (2018). *11 Facts About Teen Pregnancy*. [Online] Available at DoSomething. org: https://dosomething.org/article/11-facts-about-teen-pregnancy [Accessed 23 Oct. 2024].

Centers for Disease Control. (2021). *About Teen Pregnancy*. [Online] Available at Reproductive Health: https://www.cdc.gov/reproductive-health/teen-pregnancy/index.html [Accessed 23 Oct. 2024].

Centers for Disease Control. (2022). *Teen Birth Rate by State*. [Online] Available at National Center for Health Statistics: https://www.cdc.gov/nchs/pressroom/sosmap/teen-births/teenbirths.htm [Accessed 23 Oct. 2024].

Dweck, C. (2007). *Mindset: The New Psychology of Success*. New York: Ballantine Books.

Fessler, A. (2007). *The Girls Who Went Away: The Hidden History of Women Who Surrendered Children for Adoption in the Decades Before Roe v. Wade*. New York: Penguin.

Grigson, T. (2016). *Oklahoma's Teen Birth Rate Is Near the Highest in the Country. We Can Do Better*. [Online] Available at Oklahoma Policy Institute: https://okpolicy.org/oklahomas-teen-birth-rate-near-highest-country-can-better/ [Accessed 23 Oct. 2024].

Guldi, M. (2008). *Fertility Effects of Abortion and Birth Control Pill Access for Minors*. [Online] Available at NIH: National Library of Medicine: https://pmc.ncbi.nlm.nih.gov/articles/PMC2834388/#:~:text=Previous%20work%20has%20found%20that,among%20unmarried%20first%2Dtime%20mothers [Accessed 23 Oct. 2024].

Hofferth, S. (1987). *The Children of Teen Childbearers.* [Online] Available at NIH: National Library of Medicine: https://www.ncbi. nlm.nih.gov/books/NBK219236/#:~:text=It%20is%20clear%20t hat%20being,health%2C%20social%20and%20economic%20p roblems [Accessed 23 Oct. 2024].

International Monetary Fund. (2017). *When Women do Better, We All do Better.* [Online] Available at IMF Videos: https://www. imf.org/en/Videos/view?vid=5468229247001   [Accessed   23 Oct. 2024].

National Campaign to Prevent Teen and Unplanned Pregnancy. (2010). *Preventing Teen Pregnancy is Critical to School Completion.* [Online] Available at America's Promise Alliance: https://americas promise.org/resources/preventing-teen-pregnancy-critical-sch ool-completion/ [Accessed 23 Oct. 2024].

National Center for Education Statistics. (2024). *High School Graduation Rates.* [Online] Available at Institute of Education Sciences: https://nces.ed.gov/programs/coe/indicator/coi/high-school-graduation-rates#:~:text=In%20school%20year%202 021%E2%80%9322,cohort%20graduation%20rate%20(ACGR) [Accessed 23 Oct. 2024].

Prensky, M. (2001). Digital Native, Digital Immigrants. *On the Horizon.*[Online] Available at MarcPrensky.com: https://www. marcprensky.com/writing/Prensky%20-%20Digital%20Nati ves,%20Digital%20Immigrants%20-%20Part1.pdf [Accessed 23 Oct. 2024].

US Department of Health and Human Services. (2020). *Trends in Teen Pregnancy and Childbearing.* [Online] Retrieved from Office of Population Affairs: https://opa.hhs.gov/adolescent-health/ reproductive-health-and-teen-pregnancy/trends-teen-pregna ncy-and-childbearing [Accessed 23 Oct. 2024].

Wildsmith, E., Welti, K., Finocharo, J., & Ryberg, R. (2022). *The 30-Year Decline in Teen Birth Rates Has Accelerated Since 2010.* [Online]

Available at Child Trends: https://www.childtrends.org/publicati ons/the-30-year-decline-in-teen-birth-rates-has-accelerated-since-2010 [Accessed 23 Oct. 2024].

World Data Bank. (2022). *Population Ages 15–19, Female*. [Online] Available at World Bank Open Data: https://data.worldbank.org/ indicator/SP.POP.1519.FE.5Y [Accessed 23 Oct. 2024].

World Health Orgnaization. (2024). *Adolescent Pregnancy*. [Online] Available at World Health Organization: https://www.who.int/ news-room/fact-sheets/detail/adolescent-pregnancy [Accessed 23 Oct. 2024].

Youth.gov. (2011). *The Adverse Effects of Teen Pregnancy*. [Online] Retrieved from Youth.gov: https://youth.gov/youth-topics/ pregnancy-prevention/adverse-effects-teen-pregnancy [Accessed 23 Oct. 2024].

Recommended further reading

1. *The Girls Who Went Away* by Ann Fessler
2. *The Courage to Teach: Exploring the Inner Landscape of a Teacher's Life* by Parker Palmer
3. *Pregnant Girl: A Story of Teen Motherhood, College, and Creating a Better Future for Young Families* by Nicole Lynn Lewis
4. *Teach Like a Champion* by Doug Lemov
5. *The First-Year Teacher's Survival Guide: Ready-to-Use Strategies, Tools & Activities for Meeting the Challenges of Each School Day* by Michelle Cummings and Julia G. Thompson

About the author

Janice Airhart has been a medical technologist, biomedical research tech, freelance writer and editor, science teacher to pregnant teens, bioscience program representative, and adjunct English professor. Her memoir, *Mother of My Invention: A Motherless Daughter Memoir* won the Minerva Rising 2021 Memoir Contest and was published in 2022. Her essays and articles have appeared in *The Sun*, *The Science Teacher*, *Lutheran Woman Today*, *Concho River Review*, Story Circle Network's *Real Women Write* 2021 and 2023 anthologies, and the *Writing Strong! 35 Years of Creativity* anthology, from the San Gabriel Writers' League.

Index

www.ingramcontent.com/pod-product-compliance
Lightning Source LLC
Chambersburg PA
CBHW050349270326
41926CB00016B/3657